Wear of the Polyethylene Component Created by Rolling Motion of the Artificial Knee Joint

Markus Anton Wimmer

Wear of the Polyethylene Component Created by Rolling Motion of the Artificial Knee Joint

dem Promotionsausschuß

der Technischen Universität Hamburg-Harburg

zum Erlangen des akademischen Grades

Doktor-Ingenieur

genehmigte Dissertation von

Markus Anton Wimmer

geboren in München, Bayern

1999

1. Gutachter: Prof. Dr. sc. techn. E. Schneider

2. Gutachter: Prof. Dr.-Ing. K. Schulte

3. Gutachter: Prof. Dr.-Ing. M. Vötter

4. Gutachter: Prof. T.P. Andriacchi, Ph.D.

Vorsitzender des Prüfungsausschusses: Prof. Dr.-Ing. E. Kreuzer

Tag der mündlichen Prüfung: 25. Juni 1999

Berichte aus der Biomechanik

Markus Anton Wimmer

Wear of the Polyethylene Component Created by Rolling Motion of the Artificial Knee Joint

Shaker Verlag
Aachen 1999

Die Deutsche Bibliothek - CIP-Einheitsaufnahme

Wimmer, Markus Anton:
Wear of the Polyethylene Component Created by Rolling Motion of
the Artificial Knee Joint / Markus Anton Wimmer.
- Als Ms. gedr. - Aachen : Shaker, 1999
 (Berichte aus der Biomechanik)
 Zugl.: Hamburg-Harburg, Techn. Univ., Diss., 1999
ISBN 3-8265-6634-3

Umschlagentwurf: Peter Schmid, AO-Center, Davos

Copyright Shaker Verlag 1999
All rights reserved. No part of this publication may be reproduced, stored in a retrieval system, or transmitted, in any form or by any means, electronic, mechanical, photocopying, recording or otherwise, without the prior permission of the publishers.

Printed in Germany.

ISBN 3-8265-6634-3
ISSN 0946-3232

Shaker Verlag GmbH • P.O. Box 1290 • D-52013 Aachen
Telefon: 0049/2407/9596-0 • Telefax: 0049/2407/9596-9
Internet: www.shaker.de • eMail: info@shaker.de

Das Knie

Ein Knie geht einsam durch die Welt.
Es ist ein Knie sonst nichts!
Es ist kein Baum! Es ist kein Zelt!
Es ist ein Knie sonst nichts.

Im Kriege ward einmal ein Mann
erschossen um und um.
Das Knie allein blieb unverletzt -
als wärs ein Heiligtum.

Seitdem gehts einsam durch die Welt.
Es ist ein Knie, sonst nichts.
Es ist kein Baum, es ist kein Zelt.
Es ist ein Knie sonst nichts.

      CHRISTIAN MORGENSTERN

Meiner Familie

Preface

More than three-hundred years ago, a man called Newton was asleep under an apple tree, when suddenly an apple fell on his head. He awoke and established the Laws of Gravity. Perhaps, if Sir Isaac Newton would have slipped on a banana skin, he might have recognized the importance of friction between surfaces in relative motion and would have conceived the science of tribology. As we know, bananas were not available in Europe at this time...

This tale was told during the opening address at the EUROTRIB'89 in Helsinki, when the society president reminded the members about the founding concept of tribology in 1966 by the British Ministry of Education and Science. Tribology, derived from the Greek terms rubbing and science, has been defined as "the science and technology of interacting surfaces in relative motion". Today, tribology is no longer a neglected concept but has made its appearance in most of the scientific branches, including orthopaedics.

It was my former materials professor at the Technical University of Munich, Prof. H.M. Tensi, who strongly (but unsuccessfully) warned me not to enter the field of tribology. Several years later I can understand what he meant by "the difficulty in determining friction and wear due to the complex and interactive roles of contact mechanics, chemistry, thermo- and fluid dynamics". Nevertheless, I enjoyed my time as a research fellow in the Biomechanics Section of the Technical University Hamburg-Harburg, and I am deeply in debt to my principal advisor, Prof. E. Schneider, who gave me the freedom to pursue this area of research in his laboratory. Similarly, I have to thank my mentor in knee biomechanics, Prof. T.P. Andriacchi (Stanford University, Palo Alto), who never tired of listening to my (abstruse?) ideas and always encouraged me to continue my research. I also would like to acknowledge Prof. K. Schulte and Prof. M. Vötter who participated on the advisory board and Prof. E. Kreuzer who took the chair during my defense.

It is obvious that the amount of research presented here could not have been conducted without the help of others. Dr. J. Loos and Prof. J. Petermann from the University of Dortmund welcomed me with open arms to their facilities and taught me the principles of polymer analysis using electron microscopy. Dr. D. Drechsler from the University of Münster undertook the same efforts in terms of atomic force microscopy. Prof. A. Fischer from the Gesamthochschule Essen guided my tribological thinking and J.J. Jacobs, M.D. from the Rush Medical Center, Chicago contributed to my clinical understanding. Their unwavering support in this work was invaluable. I have to thank all my former colleagues for their help with this project, in particular Drs. T.H. Smit, and J. Feikes for their critical comments and the proof reading of the manuscript, L.M.O. Birken, A. Bluhm, R. Nassutt, and M. Vollmer for their technical support and bottomless well of ideas, and Priv.-Doz. M.M. Morlock for his conscious discussions and help in statistics.

Several students had a strong input in the form of individual thesis studies. Special thanks to M. Karlhuber, A. von Lersner, A. Reinholz, T. Schimanski, U. Schröder, T. Schwenke, J. Seebeck, and K. Sellenschloh who directly contributed to the topic. I am in debt to Prof. R.N. Natarajan who guided most of these students during their exchange program at the Rush Medical Center at Chicago. Maria, thank you for your tremendous energy to deal with the layout and figures. I am aware of your assistance in many aspects.

I could not ask for a more supportive nucleus than I found among my friends at and around the "Harburg Harbour Laboratories". What a big family! I will always remember these fruitful hours of conversation, sports and beer. I hope that we will continue our boat travels in the future with common aims and successful collaboration.

Abstract

The natural human knee is a complex and heavily loaded joint. The articulation must transmit large forces while allowing significant motion at the same time. Typically, the knee undergoes a six-degree-of-freedom movement including both rolling and sliding. For functional reasons, freedom of movement and stability are important characteristics for the artificial knee joint as well. Technically, the latter are demanding requirements because of mixed kinematics and high stresses put on the articulation. As a result, wear and fatigue of the tibial polyethylene bearing are common failure criteria.

In this study, the contact mechanics at the tibial surface of the artificial joint was parametrically analyzed using an inverse dynamics approach. The mathematical model predicted conditions following total knee replacement which can lead to tractive rolling during gait. Tractive rolling introduces shear forces at the tibio-femoral joint. The local coefficient of friction is the most important factor influencing the magnitude of the tractive force. The alternating direction of tractive force, plus the moving femoral contact, generate cyclic stresses, which accelerate fatigue failure of the tibial plateau.

Tibial components, which were removed for reasons other than polyethylene failure, were used to identify the prevalent wear mode and acting wear mechanism. The wear patterns found, correlated with the findings of the mathematical model. Furthermore, impurities – determined to be sodium, potassium and chlorine – were located throughout the material. Using a wear testing apparatus, it was shown that these inclusions work as indentors and initiate a rolling abrasion type of wear. Consequently, microploughing of the detached salt particles causes the development of striations on the polyethylene surface. The striated morphological change of the surface alters the lubrication regime between femoral condyles and tibial plateau, which increases the local coefficient of friction. The latter increases the tractive forces at the plateau and fatigue failure of the polyethylene becomes more likely.

This work provides insight into the basic structure of the tribosystem at the artificial knee joint. The knowledge of wear mode and acting wear mechanism is important in order to improve the wear characteristics of the bulk polyethylene and the design of the articulation. The elimination of the salt inclusions from the polyethylene component might be a first step of a successful strategy to reduce wear and fatigue of the tibio-femoral articulation.

Zusammenfassung

(Verschleiß der tibialen Polyethylen-Komponente durch traktive Rollvorgänge des künstlichen Kniegelenkes)

Das menschliche Kniegelenk ist ein mechanisch hoch beanspruchtes Gelenk. Die Kinematik setzt sich aus einer komplexen Überlagerung von Dreh- und Translationsbewegungen aller drei Raumachsen zusammen. Aus anatomisch-funktionellen Gründen sollte diese Kinematik auch beim Kunstgelenk erhalten bleiben, was jedoch eine für technische Gelenke untypische Relativbewegung und Materialbeanspruchung bedeutet. Letzteres spiegelt sich durch hohen Verschleiß der tibialen Polyethylen-Komponente und nicht selten durch Totalversagen der Artikulation wider.

Im Rahmen dieser Dissertation wurde zunächst die Kontaktmechanik anhand eines mathematischen Modells parametrisch analysiert. Es wurde deutlich, daß während der Gangphase des Patienten sog. traktive Rollvorgänge beim Kunstgelenk dominieren. Aufgrund der entstehenden Scherkräfte müssen diese für den Verschleißprozeß berücksichtigt werden. Insbesondere der lokale Reibungskoeffizient bestimmt die Ausprägung dieser traktiven Kräfte. So ist bei letzteren ein Vorzeichenwechsel möglich, obwohl die Richtung der Relativbewegung beibehalten wird. Diese Charakteristik - in Verbindung mit der speziellen Kontaktkinematik - induziert an einigen Stellen des tibialen Plateaus eine dreimalige Zug/ Druck Wechselbeanspruchung pro Gangzyklus, was die Ermüdung des Materials beschleunigt.

Anhand von Explantaten wurden Verschleißart und -mechanismus studiert, wobei die grundlegenden Erkenntnisse des mathematischen Modells bestätigt werden konnten. Die Entdeckung eingeschlossener Salzkristalle im Polyethylen legte die Vermutung nahe, daß der Abriebprozeß durch Kornwälzverschleiß eingeleitet und gesteuert wird, was durch ein Experiment bestätigt werden konnte. Das Pflügen dieser mikrometergroßen Salzpartikel bedingt eine Furchung der tibialen Komponente. Die damit verbundene Erhöhung des lokalen Reibungskoeffizienten führt zu einer stärkeren Ausprägung der traktiven Kräfte und somit zu einer Steigerung der Scherbeanspruchung des tibialen Plateaus.

Durch diese Arbeit wurden grundlegende Erkenntnisse über das Beanspruchungskollektiv, und den vorherrschenden Verschleißmechanismus am künstlichen Kniegelenk gewonnen. Beides sind unabdingbare Voraussetzungen für eine fundierte Verbesserung des Prothesendesigns und der verwendeten Materialien. Durch die Eliminierung der Salzeinschlüsse aus dem tibialen Werkstoff könnte die Scherbeanspruchung gesenkt und dessen Lebensdauer gesteigert werden.

Table of contents

PREFACE I

ABSTRACT (ENGLISCH AND GERMAN) III

TABLE OF CONTENTS V

NOMENCLATURE XI

ABBREVIATIONS XIII

1 INTRODUCTION 1
   1.1 Historical and Contemporary Perspective of Total Knee Arthroplasty 1
      1.1.1 Evolution of Total Knee Arthroplasty (TKA) 1
      1.1.2 Survivorship and Failure in Total Knee Arthroplasty 2
   1.2 Loading and Relative Motion at the Tibial Articulation in TKA 3
   1.3 Well-aimed Improvement of Wear Resistance at the Knee Joint 4
   1.4 Purpose and study design 5

2 BASIC SCIENCE OF THE KNEE AS RELATED TO TOTAL KNEE ARTHROPLASTY 9
   2.1 Global Anatomy of the Normal Knee 9
   2.2 Functional Knee Biomechanics 13
      2.2.1 Kinematics during Flexion and Extension of the Knee 13
      2.2.2 Dynamics of the Knee 15
   2.3 Mathematical Modeling of the Knee Joint 19
      2.3.1 Principles of Approach 19
      2.3.2 Resultant Joint Forces 20

3 PRINCIPLES OF POLYMER TRIBOLOGY 23
   3.1 The System Approach to Tribology 23
      3.1.1 History of Tribology 23
      3.1.2 System Parameters 24
   3.2 Microstructure and Wear of Polymers 26
      3.2.1 Polymer Morphology 26
      3.2.2 Structure Dependence of Mechanical Properties 27
      3.2.3 Surface Conditions 28
      3.2.4 Degradation of Polymers 29
      3.2.5 Microstructure and Properties of UHMWPE 30

3.3 Topography of Surfaces in Contact 30
    3.3.1 Measurement of Surface Topography 31
    3.3.2 Contact Mechanics 32
3.4 Friction of Polymers 34
    3.4.1 The Laws of Sliding Friction 34
    3.4.2 Theory of Friction 34
    3.4.3 Rolling Friction 36
3.5 Wear of Polymers 37
    3.5.1 Wear mechanisms 37
    3.5.2 Wear modes 39
    3.5.3 Third Bodies 40
3.6 Lubrication 41

4 SYSTEM ANALYSIS OF WEAR AT THE TIBIO-FEMORAL JOINT 43
4.1 Failure of Total Knee Arthroplasty 43
    4.1.1 Reasons for Revision 43
    4.1.2 Biologic Response to Wear Products 43
4.2 Factors Influencing Polyethylene Wear in TKA 46
    4.2.1 Material Factors 46
    4.2.2 Mechanical Factors 54

5 CONTACT MECHANICS AT THE TIBIAL SURFACE 59
5.1 Introduction 59
    5.1.1 The Role of Friction in the Analysis of Joint Dynamics 59
    5.1.2 Tractive Forces during Rolling Movement 59
    5.1.3 Purpose 61
5.2 Materials and Methods 61
    5.2.1 General Description 61
    5.2.2 Kinematic Approximation 63
    5.2.3 Approximation of Ligament and Muscle Force Vectors 64
    5.2.4 Model Input 65
    5.2.5 Stress Analysis 66
5.3 Results 68
    5.3.1 General Force Pattern for Normal Gait 68
    5.3.2 Influence of the Coefficient of Friction 70
    5.3.3 Influence of Tibial Conformity 71
    5.3.4 Influence of Gait mechanics 71
    5.3.5 Influence of Patella Position 73

    5.3.6 Stresses in the Tibial Component 73
  5.4 Discussion 75
    5.4.1 Coefficient of Friction and TKA Dynamics 75
    5.4.2 Patient and Surgical Related Factors Affecting Surface Traction 76
    5.4.3 Limitations of the Knee Model 77
    5.4.4 Tractive Rolling and its Implication on the Stress Conditions 78

6 EARLY WEAR REGIME IN RETRIEVED TKA COMPONENTS 81
  6.1 Introduction 81
    6.1.1 Retrieval Analysis of Knee Replacements 81
    6.1.2 Early Surface Changes on Tibial Components 81
    6.1.3 Purpose 82
  6.2 Materials and Methods 82
    6.2.1 Materials Selection 82
    6.2.2 Probe Preparation 84
    6.2.3 Mapping of Surface Wear 85
    6.2.4 Quantitative Analysis of the Surface Texture 86
    6.2.5 Microscopic Description of the Surface 88
    6.2.6 Determination of Impurities 89
    6.2.7 Analysis of Subsurface Characteristics 91
  6.3 Results 93
    6.3.1 Patterns of Wear on the Tibial Plateau 93
    6.3.2 Location and Area of the Striated Patterns 96
    6.3.3 Surface Texture 97
    6.3.4 Microscopic Features of the Surface 100
    6.3.5 Impurities 103
    6.3.6 Subsurface Characteristics 107
  6.4 Discussion 109
    6.4.1 Wear Pattern and Tibio-Femoral Contact Mechanics 109
    6.4.2 Wear Pattern and Changes with Time in Situ 110
    6.4.3 Relationship between Macro and Micro Surface Morphologies 111
    6.4.4 The Occurrence of Salt Impurities in UHMWPE 112
    6.4.5 The Role of the Salt Particles in the Wear Process 113
    6.4.6 The Role of the Striated Pattern in the Wear Process 114
    6.4.7 Wear History of the Tibial Plateau 115

7 DAMAGE DUE TO TRACTIVE ROLLING ON THE TIBIAL PLATEAU 117
  7.1 Introduction 117

- 7.1.1 Wear Testing of Total Knee Arthroplasty 117
- 7.1.2 Surface Damage and Contact Kinematics 118
- 7.1.3 Surface Damage and Third Bodies 119
- 7.1.4 Purpose 119
- 7.2 Methods and Materials 119
  - 7.2.1 Wear Testing Configuration 119
  - 7.2.2 Testing Protocol 124
  - 7.2.3 Test Specimens and Conditions 126
- 7.3 Results 126
  - 7.3.1 Traction Coefficient 126
  - 7.3.2 Damage Caused by Tractive Rolling 127
  - 7.3.3 Influence of Tractive Force Direction on Kinematics and Wear 129
  - 7.3.4 Micro-damage due to Detached Salt Minerals 130
- 7.4 Discussion 131
  - 7.4.1 Effect of Tractive Forces on the Kinematics 132
  - 7.4.2 Damage Increase due to Tractive Force 132
  - 7.4.3 Rolling Abrasion due to Detached Salt Minerals 133
  - 7.4.4 Limitations of the Study 134
  - 7.4.5 Transferability of Wear Patterns to Findings on Retrievals 135

8 ANALYSIS OF FRICTION WITH DIFFERENT SLIDE-ROLL RATIOS 137
- 8.1 Introduction 137
  - 8.1.1 Rolling and Sliding of Knee Prostheses 137
  - 8.1.2 Friction at the Articulation of Artificial Joints 138
  - 8.1.3 Purpose 139
- 8.2 Material and Methods 139
  - 8.2.1 Friction Test Configuration 139
  - 8.2.2 Test Specimens 144
  - 8.2.3 Test Protocol 145
- 8.3 Results 145
  - 8.3.1 Moment/ Velocity Curves for System Calibration 145
  - 8.3.2 Influence of the Slide-roll Ratio, Kinematic Mode and Velocity 147
  - 8.3.3 Influence of Polyethylene Surface Morphology 148
- 8.4 Discussion 148
  - 8.4.1 Surface Morphology Defines the Lubrication Regime 148
  - 8.4.2 Model Limitations 150
  - 8.4.3 Friction in the Mixed Lubrication Regime 150
  - 8.4.4 Effect of Friction Behaviour on TKA Kinematics 151

9 Wear Mechanism 153
9.1 Introduction 153
9.2 Rolling Abrasion due to Salt Impurities 155
9.2.1 Contact Mechanics of Particle Indentation 155
9.2.2 Load Carrying Capacity of the Salt Particles 156
9.2.3 Wear Mechanisms due to Rolling Abrasion 159
9.2.4 The Influence of Polyethylene Oxidation 160
9.2.5 A Hypothesis for the Development of the Striated Pattern 161
9.2.6 Limitations of the Third-Body Damage Model 163
9.3 Tractive Rolling due to Surface Striations 164
9.3.1 Wear Mechanism due to Tractive Rolling 164
9.3.2 Generation of Residual Strain and Stress 165
9.3.3 Surface Fatigue due to Accumulated Stress 166
9.4 Wear Mechanisms due to Spin 167
9.5 Consequences 167

10 Summary and conclusions 169

Appendices 173

References 193

Medical Terminology 217

Curriculum Vitae 219

Nomenclature

Symbol	Unit	Definition
a	m/s²	acceleration
a, b		muscle insertion points
a_c	mm	semi-contact width (Hertzian radius)
c_i	mm	initial contact point of the femoral condyles on the tibial plateau, measured from the most anterior aspect of the implant
f	mm	rolling lever
g	m/s²	acceleration of gravity
h, ρ, μ	-	standardized variables
l	m	lever-arm
m		mean of particle size distribution
m	kg	mass
n	-	number of particles in contact
r	m	radius
u, v	m/s	velocities
s, x, y, z	m	variable lengths
w	mm	compliance of two bodies in contact
$w_{ap\ (ml)}$	mm	A/P- (M/L-) width of tibial plateau
A	mm²	area
A_r	mm²	real contact area
C_0		instantaneous center of rotation
E	MPa	Young's modulus
F	N, BW	force (BW: normalized to body weight)
F_a	N	adhesion force
F_C	N	contact force
F_δ	N	deformation force
$F_{ext\ (int)}$	N, BW	external (internal) force
F_{Gast}	N, BW	force in the gastrocnemius muscle group
F_{Hams}	N, BW	force in the hamstrings muscle group
F_M	N; BW	muscle force
F_n	N, BW	normal force
F_P	N, BW	force in patellar ligament
F_Q	N, BW	force in the quadriceps tendon
F_{Quad}	N, BW	force in the quadriceps muscle group
F_t	N, BW	tractive force
F_{tis}	N, BW	force in lateral or medial soft tissue
$F_{x,\ y,\ z}$	N, BW	segmental external forces
G	MPa	elastic shear modulus
H		hardness
J	kg·m²	moment of inertia
L	mm	invariable length
M	N·m, BW×HT	moment (BW×HT: normalized to body weight times patient's height)

M_E	N·m, BW×HT	extending moment
$M_{ext\,(int)}$	N·m, BW×HT	external (internal) moment
M_t	N·m	generated moment at the brake
$M_{x,y,z}$	N·m, BW×HT	external segmental moments
N	-	total number of third bodies
P	N	partial load
R_a	μm	average roughness
R_{max}	μm	maximum peak-to-valley height
R_q	μm	root mean square roughness
R_z	μm	ten-point height (DIN 4768)
S	-	slide-roll ratio
T	K	temperature
U	V	voltage
W	N/m	load per unit width
$\alpha, \beta, \varphi, \Theta$	deg, rad	angles
α	deg	knee flexion angle
β	deg	patellar ligament angle
μ	-	coefficient of friction
μ_a	-	adhesive term of friction
μ_δ	-	deformation resistance
μ_d	-	coefficient of dynamic friction
μ_{max}	-	maximum coefficient of friction
μ_s	-	coefficient of static friction
μ_t	-	traction coefficient
δ		variable
η	Pa·s	viscosity of the lubricant
λ	μm	lubricant film thickness
σ		standard deviation
σ^*		standard deviation of a Gaussian distribution
σ_{yy}	MPa	normal tangential stress
σ_{yz}	MPa	maximum shear stress
σ_{zz}	MPa	normal contact stress
ν	-	Poisson's ratio
ω	rad/s	angular velocity
Δ		measurement uncertainty
Φ	deg	angle
Θ	deg	angle
Ψ	-	plasticity index

Abbreviations

ACL	Anterior Cruciate Ligament
AFM	Atomic Force Microscope
ADINA	Automatic Dynamic Incremental Non-Linear Analysis
A/P-	Antero-Posterior
BSE	Back-Scattered Electrons
BW	Body Weight
CLSM	Confocal Laser Scanning Microscope
CR	Cruciate Retaining
EDX	Energy Dispersive X-Ray Spectrometry
EHD	Elastohydrodynamic
EMG	Electromyography
IB	Insall-Burstein
LCL	Lateral Collateral Ligament
LVSEM	Low Voltage Scanning Electron Microscope
MCL	Medial Collateral Ligament
MG	Miller-Galante
M/L	Medio-Lateral
PCL	Posterior Cruciate Ligament
PS	Posterior Stabilizing
SE	Secondary Electrons
SEM	Scanning Electron Microscope
TEM	Transmission Electron Microscope
TKA	Total Knee Arthroplasty
UHMWPE	Ultra High Molecular Weight Polyethylene
VPSEM	Variable Pressure Scanning Electron Microscope

1 Introduction

1.1 Historical and Contemporary Perspective of Total Knee Arthroplasty

Over the last 30 years, total knee arthroplasty (TKA) has evolved to a successful procedure, providing pain relief and improved knee function for a variety of arthritic conditions. The technology of total knee replacement has changed considerably during these three decades and several reports have documented the relative success of currently used prosthetic designs [1-4]. However, the long-term perspective is still not satisfying and revision surgeries are common procedures.

1.1.1 Evolution of Total Knee Arthroplasty

The first artificial replacement of the knee was attempted more than a century ago, when Gluck tried in 1890 to implant an endoprosthesis made of ivory [5]. Subsequently, surgeons experimented with a variety of materials, including metals. It took until the end of World War II for substantial progress to be made. Chrome-nickel steels and cobalt based alloys, used in the weapons industry, proved to be tough, corrosion resistant materials for long-term implants. The first efforts in knee arthroplasty with metallic materials consisted of replacement of only one bearing surface of the joint. The outcome of these hemi-arthroplasties was somewhat variable and unpredictable, although they provided pain relief for most of the patients [6].

At that time, the first attempts at *total* knee arthroplasty were performed. Total knee arthroplasty involves the replacement of all bearing surfaces of the knee, i.e. both femoral condyles and the tibial plateau. The original total knee arthroplasties were of hinged design and had a metal-on-metal articulation. They were used with reasonable clinical outcome for nearly two decades starting in 1960 [7]. By then, the experience indicated that such systems exhibited several problems, including implant loosening, stem breakage, progressive metal wear, and infection. Primarily, the occurring difficulties were associated with the uni-axial mechanics of the hinges, interfering with the multi-axial biomechanics of the joint. In other words, a rigidly hinged joint does not account for reasonable simulation of the normal knee motion, in particular rotation. This causes inadequate stresses on all portions of the components, and fracture or loosening occurs with time.

When Gunston [8] described the first polycentric knee design to overcome these problems, efforts were also taken in developing a prosthesis which relied purely on the natural, ligamentous constraints of the knee. In 1973, Coventry *et al.* [9] reported on the first two-component total knee arthroplasty, the 'Geometric' prosthesis. It consisted of two metal femoral condyles and a tibial polyethylene plateau, attached directly to the bone by means of bone cement. Mechanical stability was provided by the ligaments (including the anterior cruciate ligament) and the surrounding soft tissue. Although the prosthesis was bicondylar, no attempt was made to duplicate the anatomy of the knee joint. A metal-on-polyethylene concept was chosen to tackle the problem of wear. The early results were promising, however, poor instrumentation and failure to restore normal knee kinematics complicated the procedure leading to problems in the long-term. Finally, in the mid-1970s the 'Total Condylar' prosthesis became available [10]. In difference to the Geometric design the femoral condyles were shaped anatomically with different radii in the distal and posterior portion of the implant. The long-term survivorship of this artificial joint was excellent with a revision rate of less than 5% at 10 years [1].

Figure 1.1: Total knee system (here: MG II) with a metal backed tibial polyethylene component

Contemporary designs of TKA still consist of two components: a bicondylar metal part for the femoral surface and a ultra-high-molecular-weight polyethylene (UHMWPE) insert for the tibial plateau (Figure 1.1) [*]. Many variations on this basic design have been developed, including the addition of metal backing and the modularity of polyethylene components. Parallel with the evolution of prosthetic design, the understanding of appropriate patient selection criteria, importance of preoperative planning for surgery, and operative technique have been developed [7].

1.1.2 Survivorship and Failure in Total Knee Arthroplasty

Based on 30,000 knees performed between 1976 and 1992, the Swedish Knee Arthroplasty Register [1,11] reported a continual improvement in prosthetic knee survival with a steady decline of complications. These promising observations have been reflected by the annual growing number of primary knee surgeries which have

---

[*]As a third constituent a polyethylene patella component can be used articulating with the femoral metal portion

already outnumbered the primary hip operations in the United States [12]. Nevertheless, revisions have been increasing to the same extent as primary surgeries (Table 1.1), and demonstrate the problem of limited durability of TKA. While better patient selection and better surgical technique helped to reduce early complications [1], the application of joint replacement to younger and more active people, plus the general increase in life expectancy, make wear and the consequences of wear to the leading cause of failure in TKA [13].

Table 1.1: Incidence of large joint arthroplasty in the United States [12]

Joint / Procedure		1993	1994	Change (%)
Hip		249,292	258,300	+3.6
	Hemi	93,634	93,439	-0.2
	Primary	128,928	137,415	+6.6
	Revision	26,730	27,446	+2.7
Knee		205,362	230,342	+12.2
	Primary	189,008	211,872	+12.1
	Revision	16,354	18,470	+12.9

Total failure of the polyethylene components (e.g. breakage and delamination) are commonly reported [13]. However, even in the absence of catastrophic failure, the release of polyethylene wear particles into the surrounding tissue is a potential cause of nonsuccess. Polyethylene debris, generated during normal function of the joint, has been associated with particle induced osteolysis (resorption of the implant surrounding bone) and subsequent implant loosening [14, 15]. Therefore, the precise understanding of the tribology[*] at the tibio-femoral articulation is essential for the improvement of TKA.

1.2 Loading and Relative Motion at the Tibial Articulation in TKA

The kinematics and dynamics of the knee joint differ from any other joint in the human body. The articulation must transmit large forces while allowing the mobility to permit normal function. Typically, during knee flexion, the femur rolls, slides and rotates over the tibial plateau [16]. The dynamic stability of the joint is dependent on a combination of passive soft tissue tension and dynamic muscular forces during function. As a result,

---

[*] the science of friction, wear and lubrication

the bearing surface of the knee is highly loaded and has to sustain considerable shear and compression forces [17].

Historically, considerable effort has been made to identify the loading and motion characteristics of the natural knee [*e.g.* 18-20]. These data have also been applied to the artificial knee, neglecting the increase in friction at the tibial articulation. "Low friction arthroplasty", a terminology which has been introduced by Charnley [21] for metal-on-polyethylene joints*, documents this negligence. In reality, the change in friction is substantial, since metal on polyethylene yields a 100-fold (!) higher frictional force than cartilage on cartilage [22]. Due to the increased friction, not only will higher shear forces be produced but the dynamics of the artificial knee will also be changed. Hence, a third condition of motion must be taken into account: in addition to free rolling and sliding, *tractive rolling* (rolling under conditions of tangential surface loads) will take place at the tibio-femoral articulation.

Tractive rolling has neither been considered as input for stress analyses nor in wear testing of the tibial polyethylene component. Most often pure sliding was chosen to test the tribological features of the component. However, it needs to be noticed that rolling motion changes the lubrication regime compared to sliding: the lubricant's film thickness (and thus, the coefficient of friction) is dependent on the relative velocity between the contacting bodies. The latter is increased during rolling motion of the joint. This influence of varying slide-roll ratios on the frictional coefficient of the artificial knee joint has not been investigated.

1.3 Well-aimed Improvement of Wear Resistance at the Knee Joint

It is useful to analyze friction and wear as a system property (Figure 1.2) rather than a material property, in order to take the multiple and inter-related factors into account [23]. Thus, the analysis of the structure of the tribosystem and the type of dynamic interaction between its elements should be the first step of a well-aimed,

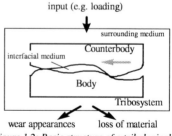

Figure 1.2: Basic structure of a tribological system

---

* The benefits of a metal-on-plastic articulation for total hip replacement were first recognized by Charnley [21] in the mid-1960s. At that time the high friction moments of metal-on-metal joints were considered as the primary reason of cup loosening.

successful strategy to reduce wear at the tibial articulation [24]. The next step involves the examination of worn parts of the tribosystem which need to be analyzed with respect to the acting wear mechanism. It should be considered that the knowledge about the effective wear mechanism is important for the improvement of the bulk material. In fact, different material or design modifications may be appropriate depending on the acting mechanism [25].

1.4 Purpose and study design

The purpose of this study is to gain a better understanding of the tribological conditions at the tibio-femoral articulation of the artificial knee joint. Knee joint loading and motion (input) as well as lubrication regime and wear mode (descriptors of the tribosystem) are evaluated. In particular, the presence of friction at the tibio-femoral articulation by the use of artificial devices is considered. The results will provide the basis to identify and discuss the underlying wear mechanism of knee prostheses made of ultra-high-molecular weight polyethylene (UHMWPE).

In contrast to technical journal bearings the loading and motion history of artificial knee joints is not easy to assess: as the *in vivo* application of this type of implant does not allow direct measurement, an indirect approach has to be taken. Mathematical analyses will help to determine the dynamic characteristics of the knee joint using inverse dynamics. Such data will be used as an input for friction and wear testing, and the morphology of the generated wear patterns will be compared to those present on retrieved components. The retrieval pattern itself may already give valuable clues as to how the particular piece of implant has been loaded. The information provided can then be used as feedback to the computer models (Figure 1.3).

This study concentrates on the analysis of the structure of the tribosystem and the type of interaction between the input parameters and systemic conditions, under the goal, to identify the effective wear mechanism. In order to achieve these objectives the outline of the thesis is based on the following chapters (Figure 1.4). The general knowledge of knee biomechanics and polymer tribology to evaluate tibial component wear is described first. A literature survey is conducted to analyze the system correlations of wear at the tibio-femoral joint. Given the complexity of this tribological system, four parallel chapters are established to finally determine the type of wear present and its acting mechanisms:

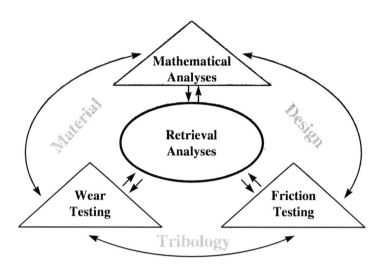

Figure 1.3: Principal elements of the research protocol. The interactive use will provide feedback information.

Analysis of the contact mechanics at the tibial surface

A mathematical model of the total knee joint is developed to analyze the occurrence of tractive forces on the tibial plateau and to test for mechanical conditions that can increase tractive forces. The significance of the tractive forces on the stresses in the bulk polyethylene are considered. In addition the impact of tractive forces on knee kinematics are studied.

Analysis of the early wear regime

An analysis of retrieved polyethylene components from TKA is conducted to gain a better understanding of the relationship between wear pattern and the suggested contact mechanics of the artificial knee. The mechanisms of the early wear process have been difficult to identify, since most retrieval studies have focused on UHMWPE failure modes. UHMWPE failure (e.g. delamination) often destroys the primary wear pattern that results during normal function and, thus, a possible basis for understanding the conditions leading to more severe damage of polyethylene. Therefore, the retrieval analysis has focused on a single design without major damage and with a geometry close to that of the computer model.

Analysis of damage due to tractive rolling

A wear testing apparatus is used to examine *defined* conditions of tractive rolling (thus, providing a known loading history) on the tibial plateau. The morphology of the generated wear pattern is compared to the patterns on the retrieved UHMWPE components.

Analysis of friction with different slide-roll ratios

A friction apparatus assists in evaluating the magnitude of friction during varying slide-roll ratios *("Schlupf")* between the femoral condyle and the tibial plateau. In order to account for the specific contact kinematics in TKA the change in surface morphology due to wear is considered.

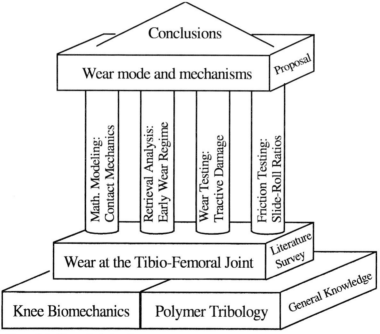

Figure 1.4: The study design (bottom up)

2 Basic Science of the Knee as Related to TKA

2.1 Global Anatomy of the Normal Knee

The knee is the largest and one of the most mechanically complex joints in the human body. The primary motions of the knee are flexion and extension. However, anteroposterior translation in the sagittal plane, internal-external rotation in the transverse plane and varus-valgus motions in the frontal plane are also important to its overall function (Figure 2.1).

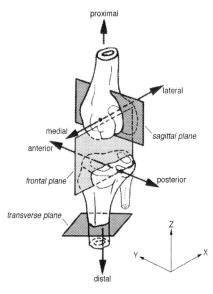

Figure 2.1: The planes of the body and some descriptive terms regarding anatomical directions. The upper (proximal) part of the knee joint displays the femur which ends in the femoral condyles. The lower (distal) part represents the tibia and fibula. Adapted and changed from [54]

Stability and freedom of movement are adjusted through the mechanistic interplay between the shape of the articulating surfaces and passive and active soft tissue structures. The following paragraphs provide a record of functional knee anatomy and is composed from the work of Kapandji [1] and Moore [2].

The distal part of the femur ends in the convex shaped, medial and lateral condyle which articulate on the flattened* medial and lateral plateau of the tibia (Figure 2.2a). The general shape of the two femoral condyles recalls – to find a technical analogy – the twin wheels of the undercarriage of an aeroplane. This specific geometry of the joint provides stability in the frontal plane while it allows freedom of motion and rotation in the remaining planes. The radii of both condyles, however, are not equal. The lateral side has a larger distal radius than the medial side, which is important for the *"screw-home"* mechanism of the knee as we will see in section 2.2.1.

---

* It should be noted that the tibial plateaus, which are separated from each other by a narrow non-articular area, are not totally flat. The medial plateau is concave, in contrast to the convex lateral.

In addition to the two tibio-femoral articulations there is a third articulation between femur and patella: the patello-femoral joint. The patella is a triangular shaped bone with its apex pointing inferiorly (Figure 2.2a). It fixed to the patellar ligament and belongs to the extensor apparatus (i.e. the tissue structures responsible for leg extension). To mention also here a technical analogy, patellar ligament and patella behave like a cable on a pulley and slide on the patellar surface. Rather simplistic speaking, both to transfer load around the corner when the knee is flexed. All bearing surfaces are covered by a layer of cartilage which insures – in combination with the synovial fluid – proper lubrication of the joint (Figure 2.2c).

Articular cartilage is a highly hydrated tissue with a relatively low compressive stiffness and permeability: these qualities allow the tissue to absorb energy and distribute loads uniformly to the underlying bone. Under normal conditions cartilage shows little or no evidence of wear and a low coefficient of friction. In pathologic situations, such as those occurring after trauma or overuse, the composition and molecular organization of it can be altered, making cartilage vulnerable to wear and degeneration[*]. As shown in Figure 2.2c the thickness of the cartilage varies over the joint surface; in areas that are highly loaded the cartilage layer tends to be thicker.

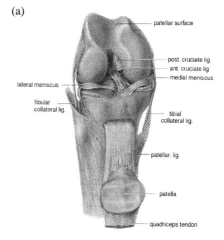

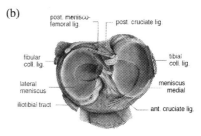

Figure 2.2:

(a) Frontal view of the right knee joint. The muscles and joint capsule are stripped and the patella is turned inferiorly.

(b) Transverse view of the tibial plateau displaying the menisci and the attachment sites of the cruciate ligaments. Adapted from [2]

---

[*] Suggested reading for more detailed information on articular cartilage is Mow et al. [53]

In the joint cavity between femur and tibia, the menisci assist the articular cartilage in distributing loads across the tibio-femoral joint. They form a set of two intra-articular, semilunar rings to accommodate for the incongruency of the articulating surfaces (Figure 2.2b). Together with the cruciate and collateral ligaments they also play an important role in stabilizing the joint. The cruciate ligaments are strong, rounded bands that cross each other similar to an X. They are named anterior and posterior according to their site of attachment on the tibia (Figure 2.2a, b). The anterior cruciate ligament (ACL), which is slack when the knee is flexed and taut when it is fully extended, prevents posterior displacement of the femur on the tibia and hyperextension. The posterior cruciate ligament (PCL), which is the stronger of the two cruciate ligaments, tightens during flexion of the knee joint and prevents anterior displacement of the femur. The fibular (lateral) collateral ligament (LCL) is a structure which serves as a primary restraint against lateral lift-off of the knee joint. Together with the tibial (medial) collateral ligament (MCL) it gives some side stability and prevents rotation of the femur medially when the leg is extended. As the collateral ligaments are slack during flexion they permit some rotation. Finally, the joint capsule, which is a strong and multi-layered structure enclosing the whole joint, has to be considered in the evaluation of passive stability. However, the primary role of the capsule is to provide nutrition to the synovial fluid and cartilage.

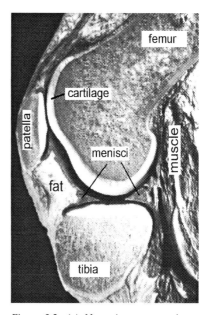

Figure 2.2: (c) Magnetic resonance image showing a healthy knee joint in the sagittal plane. Note the (white) cartilage layers which cover the femoral, tibial and patellar surfaces. Also other soft tissue structures (e.g. menisci, ligaments and muscles) can be differentiated from the bones.

The stability of the knee joint does not only depend upon the passive soft tissue structures, but also upon the strength and activity of the surrounding muscles (Figure 2.3). The most important muscle group in stabilizing the knee is the quadriceps group consisting of the rectus femoris, vastus lateralis, vastus intermedius and vastus medialis. These muscles end in the quadriceps tendon in which the patella is embedded and continues as the patellar ligament which attaches to the tibia. The quadriceps muscles are primarily responsible for the extension of the knee.

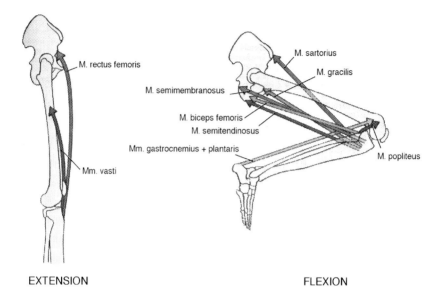

Figure 2.3: Main muscles crossing the knee joint. They are separated according to their responsible movement. Adapted from [55]

Just before full extension, the femur rotates internally (*screw-home*) and "locks" the lower extremity to a solid column which is more adapted to weight bearing. To "unlock" the knee the popliteous muscle contracts, thereby rotating the femur externally so that flexion of the knee can occur. For knee flexion the primary muscles are the hamstrings, consisting of the semimembranosus, semitendinosus and biceps femoris. The plantaris muscles and the medial and lateral heads of the gastrocnemius muscles, which originate from the distal part of the tibia, also play an important role in knee flexion. Figure 2.3 depicts the main muscles and the movements for which they are responsible.

2.2 Functional Knee Biomechanics

2.2.1 Kinematics during Flexion and Extension of the Knee

The relative motion at the knee joint can be described by three translations and three rotations which constitute the six degrees of freedom at the joint. The relative flexion-extension angle between the femur and tibia has been measured during human locomotion and found to be highly reproducible intra-individually [3]. At heel strike the knee is almost fully extended. It begins to flex, reaching a maximum of about 15°-20° during midstance. At this point the direction of the angular progression reverses and the knee fully extends again (50% of gait cycle). The joint reverses direction once more to start the pre-swing phase until toe-off occurs at approximately 63% of the gait cycle (Figure 2.4).

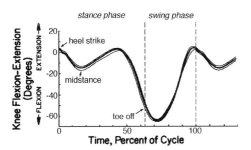

Figure 2.4: *Flexion-extension angle at the knee during the gait cycle. The heavy line represents the mean of a single subject during several trials. The thin lines indicate the standard deviation. Adapted from [56]*

Lafortune *et al.* [4] demonstrated that substantial angular and linear motions occur about all six degrees of freedom of the joint during walking. The complex motion pattern was studied using target markers that were fixed to the tibia and femur by means of intra-cortical traction pins.

Flexion of the knee progresses as a combination of rolling, sliding, and spin of the femoral condyles over the tibial plateau [1]. Experiments demonstrating this mechanism were performed as early as 1836 by the Weber brothers [5]. They evaluated the relative motion between the femoral condyles and the tibial surface by placing markers on the corresponding points of contact on both surfaces. Nearly a hundred years later it was demonstrated that the ratio of rolling to sliding varies during flexion and extension [6]. One of the models to explain the mechanism is the crossed-four-bar linkage, considered by Müller [7] and O'Connor [8,9]. In this model the insertions of both cruciate ligaments are rigidly attached to the femur and the tibia. They are represented by two crossed bars, which are not linked but held fixed firmly at their anatomical points of insertion (Figure 2.5). During flexion and extension the linkage guides the motion of the knee and results in a slide-roll ratio of approximately 2:1 during the early stages of flexion, and 4:1 during late flexion [7].

More recent experiments have demonstrated that both, the geometry of the joint surfaces *and* the soft tissue constraints determine the kinematic behaviour of the knee [10,11]. The rolling motion predominates early in flexion (0°-20°), while sliding becomes dominant at flexion angles beyond 30° (Figure 2.6). As the knee is rolling on a larger curvature laterally than medially it moves a greater distance on the lateral plateau as compared to the medial plateau. As a result, during *rollback*, the femur does not only move posteriorly but also rotates externally during flexion [12-15]. The previously mentioned *screw-home* movement is the reverse situation and gives additional stability to the knee, "more, than would be possible if the tibiofemoral joint would be a simple hinge joint" [16].

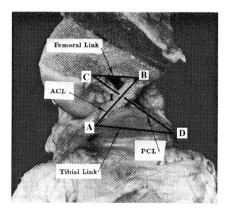

Figure 2.5: Section through a human knee with the medial femoral condyle removed exposing the cruciate ligaments and with the cruciate linkage ABCD superposed. Reproduced from [57]

The asymmetric attachment of the menisci contributes to this motion pattern. The lateral meniscus is much more mobile than the medial meniscus [1]. This is compatible with a greater movement on the lateral side than on the medial side. The more rigid attachment of the medial meniscus serves as a secondary restraint to anterior movement of the tibia providing more stability to the joint (in addition to providing increased geometric conformity) [17,18].

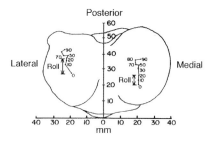

Figure 2.6: The tibio-femoral contact moves posteriorly with knee flexion. Initially, the femoral condyle rolls a greater distance laterally than medially because of its bigger radius. Beyond 20° sliding motion begins on both condyles. From [13]

The cruciate ligaments interact kinematically with the articulating surfaces [19,20]. Along the orientation of their constituent fibers they are able to resist tensile forces in the anterior or posterior directions and play a role together with the collateral ligaments in restraining varus/ valgus loading [21]. The anterior cruciate ligament usually has to be sacrificed in order to open the knee joint sufficiently and implant a total prosthesis. One of the controversies, however, is the retention or removal of the posterior cruciate ligament. An elevated cam or post, positioned between the medial and lateral bearing portion of the tibial component, restricts femoral glide off in the case of PCL removal. Moilanen and Freeman [22] argue that retention of the PCL does make the operation more complex because "balancing" the PCL often results in a PCL which is slack or too tight, causing a decreased range of motion. This elucidates that the knee works as a linkage system and a change in any link will alter the overall movement of the system. If these altered movements are incompatible with the shape of the articular surfaces, constraint forces will be generated. Thus total knee replacement which retains the PCL requires articular surfaces that closely represent the natural system, allowing the PCL to guide the knee during femoral rollback [13]. This femoral rollback has an important mechanical implication which will be discussed in more detail in the next section.

2.2.2 Dynamics of the Knee

As the femur moves posteriorly during femoral rollback, the distance from the points of contact of the femur on the tibia (Figures 2.6 and 2.7) to the line of action of the quadriceps mechanism increases [12]. This increase of the lever-arm is substantial for the efficiency of the extensor muscles (Figure 2.7a) and is reflected in total knee arthroplasty. Patients from whom the PCL was removed showed abnormal functional adaptations and had significantly more problems to climb stairs than those who had the PCL retained [23,24].

The orientation of the patellar ligament is another factor which influences the efficiency of the quadriceps mechanism. At full extension this ligament is angled between 22° and 30° anteriorly. As the knee flexes to about 60° the ligament is nearly vertical, and later in flexion it is angled posteriorly (Figure 2.7b) [25,26]. In addition to a third factor, the transfer of quadriceps force from the quadriceps tendon over the patella to the patellar ligament (Figure 2.7c) [27], those observations explain varying efficiency of the extensor mechanism with flexion. It is greatest between 15° and 25° of knee flexion and declines rapidly beyond 30° (Figure 2.8) [28]. If the lever arm of this system is reduced, more quadriceps force will be needed to balance the external forces and moments at the knee joint.

This may cause higher loading of the knee joint[*] and, thus, functional problems during daily activity [23,29].

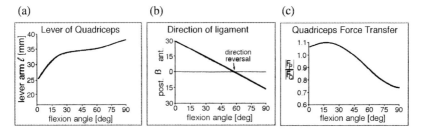

Figure 2.7: The three factors that influence the mechanical efficiency of the quadriceps with changing knee flexion. (a) change of the lever-arm 1 of the quadriceps with knee flexion; (b) change in the direction of the patellar ligament with knee flexion; (c) change in the force transfer of quadriceps tendon F_Q and patellar ligament F_P. Data from [12,25,27]

Ground reaction forces, arising during human locomotion, impose *external forces* and *external moments* at the knee joint, which must be balanced by a set of internal forces (Figure 2.9). These internal forces consist of forces generated by muscles, bone-to-bone contact forces and forces of soft tissues constraints. Out of all these structures, the muscles are in the best position to resist the external moments, because they have sufficient lever-arms, defined from their lines of action to the point of contact at the joint.

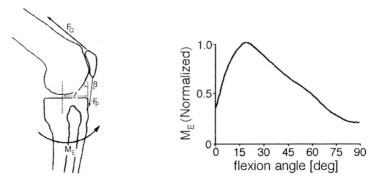

Figure 2.8: For a given constant quadriceps force, the extending moment M_E is maximum around 20° of knee flexion because of changes in the lever-arm, ligament direction, and quadriceps force transfer. Note that the plotted moment M_E is normalized to the occurring maximum. From [28]

---

[*]It should be noted that the extensor mechanism has important implications on both the loading of the tibio-femoral and patello-femoral joints. Problems with the patellar component are also common in TKA, but will not be further evaluated since the scope of this study is the mechanics of the tibio-femoral joint.

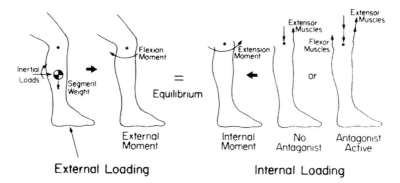

Figure 2.9: External loading of the knee has to be in equilibrium with internal loading. For example the ground reaction force plus the inertia of the limb segments may produce a moment tending to flex the knee. In order to balance this external moment a moment of equal magnitude and opposite direction has to be produced by internal structures. This "net moment" can be generated either with or without antagonists. Adapted from [36]

The external moment patterns during stance and swing phase of gait are shown in Figure 2.10. Typically, at heel strike, there is an external flexion-extension moment tending to extend the knee joint. In order to balance this moment internally, the flexor muscles have to become active. As the knee moves into mid-stance, the external moment reverses direction, demanding the action of the extensor muscles. The external moment reverses direction again during late midstance, activating the flexor muscles. Finally, at toe-off the extensors have to become active once more [30].

The muscular activity pattern during normal gait can be obtained using electromyography (EMG). EMG signals, however, give no direct access to the generated muscle force and the produced moment [31-33]. EMG can be used, however, to determine which muscle is active and to what extent. The electromyographic activity of the flexor and extensor group during level walking is shown above the flexion-extension moment graph (Figure 2.10). The activity pattern does not follow simple mechanical rules: the extensor muscles are active although there is an external extending moment, and flexor muscles work in spite of an occurring external flexing moment. This phenomenon is referred to as *antagonistic muscle activity* and increases the stability of the joint [34].

The measured flexion-extension moments during gait can, therefore, only be interpreted in terms of "net quadriceps" and "net flexor" demand, while EMG is a valuable tool in defining whether the muscle is "on" or "off"; e.g. from the activity pattern of Figure 2.10 it can be deduced that at heel-strike it is primarily the muscles of the hamstring group which are active to balance the extension moment, while during mid-stance the extension moment is accommodated by the medial and lateral gastrocnemius. The capability of muscles for *synergistic activity* (several muscles providing the same biomechanical function) and the presence of antagonistic muscle activity during human locomotion makes the analysis of the effective load on the joint complex (see chapter 5).

When the knee is near full extension, the patellar ligament is anteriorly angled (Figure 2.7). Due to this orientation of the ligament, the quadriceps muscles pull the tibia forward against the resistance of the anterior cruciate ligament. Patients who no longer have an ACL usually adapt by avoiding the normal use of the quadriceps muscles [35,36]. The patients achieve this by changing their upper body position in such a way that the vector of the ground reaction force passes the knee anteriorly, causing net flexor demand (Figure 2.11). Due to this adaption the patients protect their collateral ligaments and medial meniscus from stretching. A similar gait pattern has been recognized in patients with total knee prostheses (and sacrificed ACL), although not to the same extent (50% vs. 75%) [24,37]. Its influence on the contact mechanics of an artificial knee implant will be investigated in detail in chapter 5.

During walking the associated adduction moment (Figure 2.10) produces an asymmetric load distribution in the frontal joint plane and forces the knee into varus (*"O-Beinstellung"*). This can be related to the vector of the ground reaction force while typically passes the knee medially.

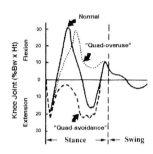
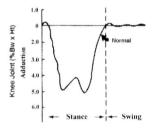
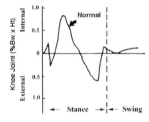

Figure 2.10: The moments at the knee joint during level walking along with myoelectric activity of various muscle groups. The flexion-extension moment, abduction-adduction moment, and internal-external moment act in the sagittal, frontal, and transverse plane respectively (Figure 2.1). All moments are normalized to body-weight × height. Note the occurrence of abnormal flexion-extension patterns which are frequently observed in patients having a total knee replacement. Adapted from [3]

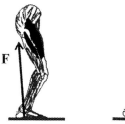
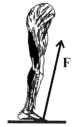

Figure 2.11: *The line of action of the ground reaction force requires either "net extensor" or "net flexor" demand. The patient is capable to control the force direction with movement of the upper body.*

The collateral ligaments help to balance the loads between the medial and lateral condyles of the knee. Proper tensioning of the lateral collateral ligament is therefore absolutely critical to prevent lift-off of the lateral condyle in TKA [38]. Also the risk of leaving the knee in residual varus is evident: residual varus results in increased adducting moments and, thus, higher medial compartment loads leading to subsequent failure of the implant [13].

2.3 Mathematical Modeling of the Knee Joint

2.3.1 Principles of Approach

Contact forces occurring at the articulation of the knee are of significant biomechanical interest for both the understanding of the evolution of joint degeneration and the design of total joint replacement [39]. It has been difficult to conduct contact force measurements at the human knee, whereas such measurements have been performed at the hip by several investigators [40-43]. The more complex behaviour of the knee in terms of both kinematics and loading makes it difficult to use instrumented prostheses (although there have been recent attempts [44]). To obtain insight into the dynamics of the joint, mathematical modeling is still the method of choice.

There are only very few models which have taken a direct dynamics approach. Examples are the models of Wismans, Andriacchi and Blankevoort [20,45,46]. In the direct dynamics approach the forces are causes and motions are effects, in contrast to the inverse dynamics approach where the equilibrium position is given (and the forces are recalculated). The difficulty inherent in the direct dynamic models is that they have to deal with the exact mathematical description of the incorporated structures in order to determine the correct displacements. Thus, in the past, most knee joint models have used the inverse dynamics approach rather than the direct dynamics approach.

In the inverse dynamics approach the *external* forces and moments are calculated from measurement of the three-dimensional position of the limb segments and the ground reaction force. The body segments are approximated as a system of rigid links connected by movable joints [30]. It is then assumed that the external forces and moments must be balanced by a set of forces and moments acting internally, which are primarily generated by muscle contraction, other soft tissue tension and articular reaction forces (see Figure 2.9). Due to the redundancy of the internal structures, this approach induces more unknowns than can be solved with the number of equations available.

In general two attempts have been made to solve this indeterminate problem. The first reduces the unknowns by grouping the muscles and other soft tissue structures into functional units. The second uses an optimization criterion to solve the supernumerary mechanical equations. The optimization criterion works based on to either trying to maximize or minimize a physical parameter (e.g. muscle endurance) which is modeled as an objective function.

2.3.2 Resultant Joint Forces

Morrison [47,48] was the first to report a method to calculate contact loads at the tibio-femoral articulation. He grouped the muscles acting at the knee into the hamstrings, gastrocnemius and quadriceps and divided the ligaments into the cruciate and collateral. Using this method he was able to turn the problem into a three-dimensional statically determinate one. The general characteristic of the calculated tibio-femoral contact force during stance phase of level walking showed three peaks with a maximum as high as 3 times body weight (Figure 2.12).

Seireg and Arvikar [49] addressed the indeterminate nature of the problem using an optimization criterion. They modeled the lower extremities as a system of seven segments connected by 31 muscles. Together with the joint reaction components of each plane the model yielded 104 unkowns by 42 given equilibrium equations. In order to solve the problem they established a linear objective function which had to be satisfied by all equations: the sum of the muscle forces plus a weighted sum of left-over joint moments (which were assumed to be taken by the ligaments) had to be minimized. The calculated tibio-femoral contact force showed a similar pattern to that found by Morrison, but with its maximum force exceeding 7 times body weight (Figure 2.12).

The difficulty with optimization methods has been to define the relevant physiological criterion. This problem has been discussed by several authors who tried to evaluate the most feasible criterion for human locomotion [50,51]. The incomplete information about the physiological function and the role of the individual muscles, in addition to mathematical simplifications of the anatomy make the calculation of the joint contact force disputable. Still, the exact physiological function and role of synergistic and antagonistic muscles remains unclear. Schipplein *et al.* [38], investigating this influence of antagonistic muscle activity on contact loads parametrically, reported forces between 3 and 5 times body weight during the stance phase of gait (Figure 2.12).

Unlike other studies, this relatively simple approach from Schipplein *et al.* points to the importance of antagonistic muscle activity for joint stabilization. In contrast it is rather challenging to define an objective function considering muscle co-contraction. The problem then becomes even further complex when physiological and mechanical aspects are incorporated into the model. For example, the maximum force that can be physiologically generated by a muscle depends on its length and on the velocity of contraction* which makes the optimization process dependent on patient activity or pathological conditions. In conclusion, a parametric model provides a rough but suitable approach for the estimation of resultant joint forces.

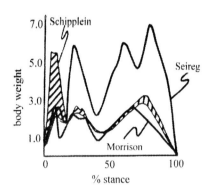

Figure 2.12: The resultant joint forces (bone-to-bone contact forces) at the normal knee joint from three different authors. The forces - normalized to body weight - are plotted against the stance phase of gait and have been calculated by solving the indeterminate problem by three different techniques. Data from [38,49,58]

---

* The relationship between muscle force, length and velocity has been described by Hill [52]: at low velocities of contraction, the muscle can provide high forces. At high velocities, its potential to generate force diminishes.

3 Principles of Polymer Tribology

3.1 The System Approach to Tribology

3.1.1 History of Tribology

For several millenniums, human mankind has been aware of friction and wear. Cave men used friction to light a fire by rubbing sticks on a piece of wood. The Egyptians and Sumerers (3500 - 35 BC) used leather belts to reduce friction between axle and wheel of their carriages [1]. Water, oil, fat or bitumen served as early lubricants and it was the use of these materials which permitted the transport of the stone blocks needed to build up pyramids. The first scientific approaches to analyze friction and wear were made by Leonardo da Vinci (1452-1519). Leonardo measured the frictional forces of bodies sliding on horizontal and inclined planes. He found that the friction force depended on the normal load but was independent of the apparent contact area. Leonardo studied particularly the wear on technical bearings and recommended an alloy of three parts of copper and seven parts of tin as material of choice.

Two centuries later, Amontons (1663-1705) confirmed independently of the work of Leonardo da Vinci that the friction force depends on the normal load but not on the apparent area of contact. His analyses, in addition to Coulomb's (1736-1806) observations on dry friction of solid bodies, are commonly referred to as Coulomb's laws. Newton (1646-1727) was also active in this scientific branch. He was aware of the favourable influence of lubricants in reducing friction and wear and described their viscosity by means of fluid mechanics.

Nearly another two hundred years later, Stribeck (1861-1950) determined adhesion, deformation and lubrication as the principal portions of friction and formulated his laws. It was then possible to describe the sliding frictional behaviour of both lubricated and unlubricated metal-on-metal joints. Because of the complex nature of friction and wear and the required interdisciplinary approach, a commission was founded by the British Department of Education and Science in 1966, with the task of defining the scientific evaluation of friction, wear and lubrication and of introducing the necessary interdisciplinary tools needed to solve the related problems [2].

In contrast to natural synovial joints, which have excellent characteristics of low friction, high load carrying, shock absorption, and long endurance, many total joint prostheses have shown serious problems such as component failure and joint loosening due to wear. It was Charnley who introduced tribological thinking to the medical community because of his early, disastrous experience with Teflon® [28]. This low friction material had extremely poor wear characteristics when articulating against a metallic counterbody. Voluminous amounts of generated debris caused a severe inflammatory response which rapidly led to loosening of the artificial devices. This emphasized the need to consider all the characteristics of a material carefully, prior to the implantation in humans.

3.1.2 System Parameters

Friction is the resistance to motion and arises from the interaction between solids at the real area of contact [3]. Friction and wear is a serious cause of energy and material dissipation. While the energy is usually dissipated as heat, the material is lost as particulate debris from the bulk and is referred to as wear. In DIN 50320 [4] wear is defined as "the progressive loss of material from the surface of a solid body due to mechanical action". While fracture of the component may also occur in extreme cases, wear usually results in dimensional changes and surface damage. This can cause secondary problems such as misalignment or insufficient motion of the joint. The generation of wear debris, however, is in many cases even more serious than the actual dimensional change of components. Entrapped particles act as interfacial material and may change the acting wear mechanism. For example, it has been shown that generated wear particles have load carrying capacity if they exceed a certain volume [5]. Thus, contact conditions between the two rubbing surfaces are completely changed if these generated particles are entrapped. The history of entrapped particles is therefore as important as their generation.

While the mechanical properties of engineering materials can be described in terms of yield strength, Young's modulus, fracture toughness, etc., friction and wear are not intrinsic material properties but are characteristics of the system [6]. The structure of such a tribological system consists of four principal elements: solid body, counterbody, interfacial medium and environment [7]. The input to the system are the 'operating variables' *("Beanspruchungskollektiv")* while the output can be defined as loss of energy and material (Figure 3.1).

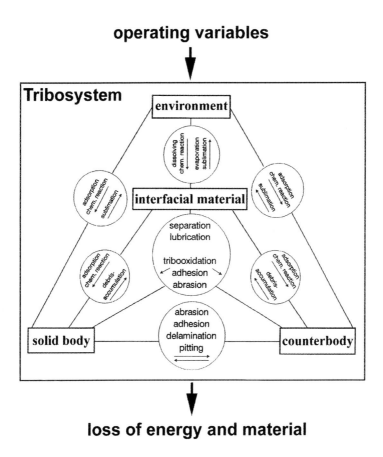

Figure 3.1: General description of the tribosystem which consists of four principal elements: the two bodies in contact, the interfacial material, and the surrounding media. All these elements can affect each other and change the mechanism of interaction. For example interfacial material can scratch the counterbody in a way that the adhesive interaction between solid and counterbody is turned into abrasion. Reproduced from [7]

Because of the complex nature of friction and wear, the problems of tribology are difficult to assess by a simple model. In order to follow the system theory of Figure 3.1, it would be necessary to analyze the wearing specimens under real conditions, i.e. incorporated into the concrete machinery under defined operating conditions. Of course this is very time consuming and expensive. Therefore, Uetz *et al.* [8,9] suggest a step by step analysis, whereby the models of wear approximate reality more closely with each step (Figure 3.2).

Fischer [10] stresses the importance of knowledge of the acting wear mechanisms*, since it is known that even for the same wear mode*, different design or material modifications may be appropriate. For example, the parameters of the wear mode 'rolling abrasion' (*"Wälzverschleiß"*) of a metal-on-metal bearing can be changed in such a way that either the mechanisms 'adhesion' (mostly plastic interaction) or 'surface fatigue' (mostly elastic interaction) apply. Thus, a successful plan to improve the bulk characteristics of the materials in contact demands an exact understanding of the structure of the tribosystem.

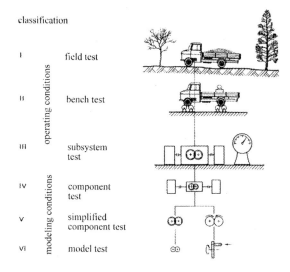

Figure 3.2: Classification of different types of tribological testing. Adapted from [8]

3.2 Microstructure and Wear of Polymers

3.2.1 Polymer Morphology

The microstructure of materials is an important parameter for the tribology of polymers. For example the length of the macromolecules of ultra-high-molecular-weight polyethylene (UHMWPE) was one of the reasons to choose this material for total joint replacement. The macromolecules of polymers, which result from linking of 10^3 to 10^5 monomer units by covalent bonds, can be linear, branching, slightly cross-linked, or cross-linked to a network. Linear macromolecules can be arranged like a pad of cotton wool or in a regular array; those zones are called amorphous or crystalline respectively (Figure 3.3). Polymers consisting of linear macromolecules are called

---

* Definitions see sections 3.5.1 and 3.5.2

thermoplastics, in contrast to polymers with a three-dimensional, cross-linked network which are called duromeres. The latter can not be remolded by heating.

The strength of a single macromolecule is determined by the covalent bonding of the carbon atoms. Secondary bonds between the macromolecular chains of linear polymers are due to van der Waals

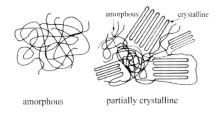

amorphous   partially crystalline

Figure 3.3: Molecular arrangement of amorphous and partially crystalline polymers

forces, while cross-links between the chains of duromeres are formed by covalent bonds. The relative freedom of linear polymers allows different arrangements, either amorphous or partially crystalline and defines a supermolecular structure (Figure 3.4). Small crystallites are formed by folded chains (referred to as lamellae) or extended chains (referred to as fibrils).

The crystallites may be arranged as spherulities, which consist of lamellar structures and amorphous regions in-between. Also other morphologies of crystal arrangement are possible, like fiber-, needle-, or shish-kebab structures, which are dependent on the circumstances of solidification after processing [11].

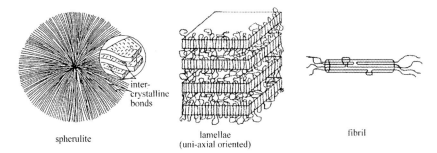

Figure 3.4: Frequently occurring super-molecular structures of polymers. Reproduced from [11]

3.2.2 Structure Dependence of Mechanical Properties

As expected the physical properties of polymers are affected by morphology. This is true for both the molecular and the supermolecular level. The chain length of the macromolecules (or molecular weight) determines the elastic modulus as well as the overall strength of the polymer. It is also an important factor for the formation of particulate debris: the longer the macromolecular chains, the more difficult to separate them from the bulk [12]. A polymer with a needle structure has a much higher tensile strength as compared to polymers with a fiber structure. The yield strength is influenced by crystallinity, but more or less independent of chain length. As the degree of crystallinity increases, both Young's modulus and yield strength increase, but ductility may decline. Similar to metals, sufficient ductility is necessary for good wear properties, otherwise brittle fracture may occur [3]. Therefore, in addition to chain length, the degree of crystallinity is an important material property for wear resistance of (semi-crystalline) polymers and can range from 10 to 90% [11].

The mechanical properties of polymers show a strong dependence on test temperature and loading rate. In particular, the wear of polymers may be affected by temperature because of their low thermal conductivity combined with relatively low melting temperatures. Depending on the morphology or crystallinity of polymers their yield strength, hardness and elastic modulus decrease with increasing temperature [3].

3.2.3 Surface Conditions

Since friction and wear arise from conditions at the interface, the properties of materials at the surface need to be looked at. The hardness of a material, which characterizes its ability to resist plastic deformation at the surface, is an important design parameter for wear resistance [12]. The hardness of polymers usually increases with increasing crystallinity and chain length, as well as decreasing free volume [3]. The free volume declines with cross-linking. When different materials are paired in bearings, the softer one tends to be transferred to the harder one during wear. This polymer transfer may be reduced to some degree by altering the hardness of one or the other material [12].

Polymers show marked differences in surface properties compared to metals, arising from the fact that the surfaces of metals react strongly with oxygen to generate oxide layers, and that the metal surfaces bond with molecules of the environment due to their high surface energies. Consequently, wide variations in the frictional properties of metals are found depending on the cleanliness of their surfaces. The occurrence of contamination is less important for polymers [13]. It has to be considered, however, that many polymers show degradation phenomena in a reactive environment, as for example with body fluids.

3.2.4 Degradation of Polymers

Polymers adsorb low-weight molecules *in vivo*, usually water and lipids. At the bulk level, these molecules cause swelling and weight gain [12]. At the molecular level, chain scission and cross-linking may occur, either due to free radical depolymerization or due hydrolysis [14]. The latter is accelerated by sterilization and subsequent loading of the implant, as has been reported in particular for UHMWPE (Figure 3.5).

When gamma irradiated at the time of manufacturing, retrieved UHMWPE articular components show increasing extent of oxidation with time (measured over months or years). These oxidized regions correspond to regions with increased density and crystallinity and become extremely brittle after a few years. Due to continuous loading of the implant cracks may be initiated resulting in increased wear of the articulation [34].

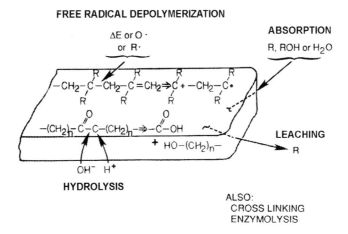

Figure 3.5: Degradation processes of polyethylene in a reactive environment. The presence of oxygen (which acts as a radical) increases the extent of carbonyl groups. The latter can be quantitatively measured using infrared spectrometry. Adapted from [12]

Mechano-chemical phenomena in polymers are well documented but still not completely understood. It is thought that strain imposed on the polymer chains causes degradation, both directly through mechanical attack and indirectly by accelerating chemical attack (e.g. through thermo-oxidation). As reported in [15] the PETERLIN model can be used to explain this strain related damage mechanism for partially crystalline polymers (Figure 3.6).

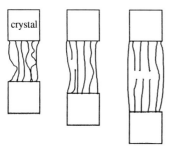

Figure 3.6: *The PETERLIN model for the fracture in partially crystalline polymers. The crystallites are connected by tie molecules or fibrils. During displacement of the crystallites the tie molecules exceed their maximum length and break.*

The crystals are connected by tie molecules or fibrils. During gradual displacement of the crystals, the shortest tie molecule exceeds its maximum possible length. The molecule breaks apart and a pair of mechano-radicals is produced. With increasing deformation, the concentration of the mechano-radicals increases. After unloading, the formation of new radicals is stopped until the strain of the sample exceeds the maximum strain of the proceeding deformation.

3.2.5 Microstructure and Properties of UHMWPE

Ultra-high-molecular-weight polyethylene (UHMWPE) is a high density polyethylene with a molecular weight of above 3×10^6 according to ASTM D 4020. About 10,000 ethylene molecules join together to form the polyethylene molecule. The polymer grows on a catalyst particle (made from titanium chloride and organo- aluminum compounds) to subsequently enclose it [29]. The resulting polymer particle is then about 10-50 times bigger than the catalyst particle, having a diameter of 50 to 500 µm. It may be further processed through compression-molding and ram-extrusion which ends in material densities ranging from 0.93 to 0.96 g/cm³ [30].

Some of the advantageous properties of UHMWPE are the abrasion resistance higher than that of any other thermoplastic material, a low coefficient of friction, the highest impact toughness of any plastic (even at low temperatures), satisfactory corrosion and degradation resistance, and resistance to cyclic fatigue. Preferred applications of UHMWPE are in chemical processing, electric devices (such as highly efficient battery separators), textiles (as fiber material) and, last not least, in total joint arthroplasty [31].

3.3 Topography of Surfaces in Contact

Engineering surfaces are far from being ideally smooth, and exhibit varying degrees of roughness. Figure 3.7 shows the difference between apparent and real area in contact. The irregularities of the surface usually consist of broad based hills with angles of inclinations of less than 15° from the base [3]. This has to be born in mind when evaluating recorded surface profiles, where those irregularities appear as sharp peaks due to the differences in vertical and horizontal magnification. Different types of experiments have shown that the variations between apparent and real areas of contact of two flat solid surfaces pressed together can be enormous. The ratio of real to apparent area of contact might be as low as 10^{-4} and depends on the distribution of irregularities, contact force and yield strength of the softer material [3].

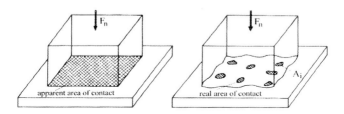

Figure 3.7: Apparent and real area of contact. Adapted from [3]

3.3.1 Measurement of Surface Topography

Surface topography is most commonly assessed using a stylus or laser profilometer, whereby the stylus or the focused laser travels over the surface and records the vertical displacement between peaks and valleys. Hence, the surface is analyzed in lines, not over an area. Parallel traces have to be performed in order to mirror the area of interest. Limitations of the stylus profilometer due to the finite dimensions of the tip radius (minimum 5 µm) have to be recognized. Also, compliant or delicate surfaces, such as those of polymers, may be distorted or damaged by the load on the stylus. Therefore, optical methods of surface measurements are attractive for such applications. In addition to laser profilometry, which ideally has a z-resolution of 50 nm, recent advances in white light interferometry have been made to record topographical features of the surface. Interference between two beams of light creates fringes which are digitally recorded. Known displacements of the reference surface cause changes in the fringe pattern from which the distribution of surface heights can be computed. The z-resolution of the white light interferometer is quite similar to that of the laser-profilometer, however, its lateral resolution is higher[*].

The most commonly quoted measure of surface roughness is the *average roughness* R_a. It is defined as the arithmetic mean deviation of the surface height from the mean line through the profile:

---

[*] see sections 6.2.4 and 6.2.5 for more details on surface profilometry.

$$R_a = \frac{1}{L}\int_0^L |y(x)| \cdot dx \tag{3.1}$$

where y is the height of the surface above the mean line at a distance x from the origin, and L is the overall length of the profile under examination (Figure 3.8).

The r.m.s. *roughness* R_q is another often used roughness measure. It is defined as the root mean square deviation of the profile from the mean line and is $1.25 \cdot R_a$ for a Gaussian distribution of surface heights. The *maximum peak-to-valley height* within the sampling length L is expressed as R_{max}. In order to reduce the effect of odd scratches or spurious irregularities the five highest peaks and five lowest valleys (measured from the mean line) are averaged as ten-point height R_z (Figure 3.8).

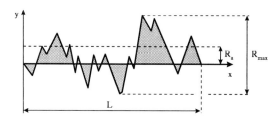

Figure 3.8: Illustration showing the surface height, y, relative to a mean line, plotted against distance. The mean line is defined so that equal areas of the profile lie above and below it. The overall length of the profile under examination is L. The parameter R_a is determined as the average of all deviations from the mean line of the surface profile; it is equivalent to the sum of the shaded areas. R_{max} is the distance from the highest peak to the lowest valley.

It is inevitable that in attempting to describe a profile by a single number, some important information about the surface topography will be lost. The above roughness measures, for example, give no information on the form and spacing of the surface irregularities. Thus, for a fuller description of the surface topography, information is needed about the *shape* and *spatial distribution* of peaks and valleys across the surface [16].

3.3.2 Contact Mechanics

As pointed out above, initial contact will occur only at a few points when two flat surfaces are brought together. These points, which represent elevated areas of the surface, are called asperities. As the normal load increases, the surfaces move closer together and more asperities come into contact. Their type of deformation defines the properties of friction and wear, which can be either elastic or plastic. Thus, the real contact area A_r is related to the total load F_n exactly in the same manner as the individual contact spots for each asperity. The following equations can be established (for uniform asperities of a single radius and height):

$$A_r \propto F_n^{2/3} \qquad (3.2)$$

for the case of purely elastic contact, and

$$A_r \propto F_n \qquad (3.3)$$

for perfectly plastic behaviour, assuming Hertz's elastic law or Meyer's indentation hardness law respectively [17]. Greenwood and Williamson [18] proposed a plasticity index Ψ which describes the transition from elastic to plastic deformation of surface asperities.

$$\Psi = \frac{E'}{H} \cdot \sqrt{\frac{\sigma^*}{r}} \qquad (3.4)$$

with

$$E' = \frac{E_1 E_2}{E_1(1-v_2^2) + E_2(1-v_1^2)} \qquad (3.5)$$

where H is the hardness of the softer material, E_1 and E_2 the Young's moduli and v_1 and v_2 the Poisson's ratios of the two bodies in contact, r the radius of the asperity hills which is assumed to be the same for all asperities and σ^* the standard deviation of a Gaussian distribution of the asperity heights [3]. While for metal surfaces the majority of the asperities – even under the lightest loads – will be plastically deformed (i.e. $\Psi > 0.6$; Figure 3.9), asperities of polymers usually react elastically (i.e. $\Psi < 1$) since their ratio E'/H is usually less than 10 while metals reach 100 and more [17].

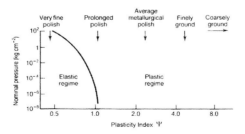

Figure 3.9: The dependence of the asperity deformation mode on the plasticity index Ψ for aluminum surfaces with different roughness. Adapted from [35]

From the above it follows that the deformation of asperities is mainly determined by the characteristics of the surface texture, hardness and elastic properties, while the applied load or surface pressure do not directly influence the transition from elastic to plastic contact.

3.4 Friction of Polymers

3.4.1 The Laws of Sliding Friction

The empirical "laws" of (dry) sliding friction are often attributed to Amontons who rediscovered them in 1699, two hundred years after Leonardo da Vinci described them for the first time:

- the friction force is proportional to the normal load;
- the friction force is independent of the apparent area of contact;

To these, a third law was added by Coulomb (1785):

- the friction force is independent of the sliding velocity.

Although most metals, and many other materials, obey the first and second law well, polymers often do not [19]. They rather show a non-linear behaviour upon load and contact area. Also the third law is not well stated. It is commonly observed that the frictional force needed to initiate sliding is usually greater than the force necessary to maintain sliding movement. Thus, the *coefficient of static friction* (μ_s) is larger than the *coefficient of dynamic friction* (μ_d). However, once motion is initiated, for many systems μ_d is found to be nearly independent of sliding velocity [17].

3.4.2 Theory of Friction

The friction of polymers is generally attributed to two sources: a *deformation* term F_δ, involving dissipation of energy, and an *adhesion* term F_a, originating from shear strength of the interface between the body and the counterbody.

$$F_t = F_\delta + F_a \qquad (3.6)$$

The deformation component of friction results from the interlocking characteristics of the two counterbodies in the tangential direction. In order to overcome this type of resistance, the material has to be deformed (or fractured), and thus work has to be carried out. During the cyclic process of deformation and recovery of material elements, energy will be dissipated as heat if the material is viscoelastic or plastic[*]. The arising frictional force is then determined by the energy dissipated. Since the deformation resistance μ_δ will be dependent on the depth of penetration, which itself depends on the material properties and normal load, it can be shown that

$$\mu_\delta \propto \left(\frac{F_n}{E}\right)^{\frac{1}{2}} \tan \delta \quad (3.7)$$

where F_n is the normal load, E the Young's modulus of the softer material, and $\tan\delta$ a mechanical loss-factor describing the damping of the material. According to Uetz [9], the above equation is valid for cylinders and spheres when $\tan\delta < 0.2$ and sliding velocities are moderate.

When two surfaces are brought into contact under load, local welding at the tips of major asperities of the surfaces brings about the adhesive term of friction μ_a. In order to shear these welded junctions a tangential force is necessary to initiate or maintain sliding movement. This idea was first published by Bowden and co-workers [20] and helped tremendously to explain the sliding friction between elastic materials. It can be shown that

$$\mu_a \propto A_r \tau_s \quad (3.8)$$

where A_r is the real contact area and τ_s is the average shear strength over the real area of contact.

For sliding friction between very smooth surfaces $F_t \approx F_a$, since F_δ is small compared to F_a. Using equations 3.2 and 3.8,

$$\mu \propto F_n^{-1/3} \quad (3.9)$$

when the polymer is perfectly elastic. Thus, for polymers, the coefficient of friction is *not* independent of the load. In reality there is fairly good agreement with equation 3.9 [19]. Remaining differences can be attributed to the presence of some plastic flow around the contacting asperities and their specific contact mechanics. In the case of purely plastic contact[†], however, μ is independent of F_n [19].

---

[*]If the material is ideally elastic no energy will be dissipated and, thus, no friction will arise due to deformation.
[†]Purely plastic contact is mainly confined to metals

3.4.3 Rolling Friction

The deformation component of friction becomes important when a hard sphere or cylinder is rolling over a viscoelastic surface*. As the sphere in Figure 3.10 rolls from left to right, a single element of the polymer in its path becomes sequentially deformed by shear and compression, while behind the sphere it recovers its undeformed shape. Because of the material's viscoelasticity, energy will be dissipated during this cycle as heat.

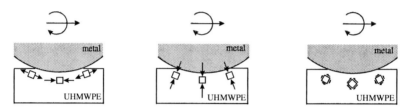

Figure 3.10: When a metal sphere indents the polyethylene surface, complex stress distributions result: (a) tensile and compressive stresses tangential to the surface, (b) compressive contact stress perpendicular to the surface, and (c) shear stresses.

Reynolds [21] already recognized that with rolling of a rigid body on a deformable base, rolling must be accompanied by sliding since after a complete revolution, the displacement of the roller center is somewhat less than the length of its circumference. The normal load produces a contact over a finite area in which some contacting points may 'slip' while others may 'stick' during rolling contact. Thus, both the deformation of the polyethylene and the slip contribute to rolling resistance in such a manner that ideal rolling contact without any energy dissipation could never be achieved [22]. Resistance to rolling is manifested as a moment of the normal load around the point of contact, in which the lever arm is referred to as the *rolling lever* (Figure 3.11). Usually, the friction of a free rolling cylinder is small and the rolling lever hardly exceeds the order of magnitude of $10^{-4} \times$ radius of contact area. However, under conditions of tractive rolling (i.e. rolling with tangential force transfer) it might rise to considerable values since the micro-slip at the interface increases [23].

With repeated loading and unloading of the polymer, energy dissipation is not only confined to viscoelasticity but also plasticity. This process takes place when the polymer

---

*It should be noted that UHMWPE has non-linear viscoelastic properties [27]

becomes deformed due to plastic flow and is known as *shakedown*. While during the first pass of the roller the elastic limit is exceeded and plastic deformation takes place, a steady state limit of deformation might still be reached. The plastic deformation introduces residual stresses, which make the contact quasi-elastic and contribute to work hardening of the material. Thus, yielding is less likely on the following cycles [23].

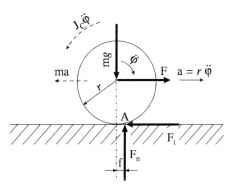

Figure 3.11: Resistance to rolling, manifested as a moment of the normal load around the point of contact

3.5 Wear of Polymers

3.5.1 Wear mechanisms

Wear is a direct result of the occurring mechanisms of friction at the interface which influence the physical and chemical reactions of the tribosystem. There are four basic mechanisms of wear, each of which obeys its own laws. According to DIN 50320 the four basic wear mechanisms are:

Adhesion

Adhesion occurs when two smooth bodies slide over each other and fragments of one surface are pulled out and adhere to the other. Later these fragments may come off the surface on to which they adhered and be transferred back to the original surface or they may form loose wear particles. The mechanism leads to the formation of local junctions between the surfaces which may be adhesive or cohesive (Figure 3.12a). While adhesion is effective between the surfaces of different materials, cohesion occurs when similar materials are welded together.

Abrasion

Abrasive wear occurs when a rough hard surface, or a soft surface containing hard particles slides on a softer surface and ploughs or cuts grooves in it. The material removed from the grooves forms the particulate debris. The micro-mechanisms due to abrasion according to Zum Gahr [3] are shown in Figure 3.12b. Microcutting removes the material due to a single pass of one abrasive asperity or particle, while microploughing displaces the material sideways. However, successive or simultaneous microploughing with many protuberances involved, causes low cycle fatigue and eventually material break-off. In addition, microploughing and microcutting introduce large strains into the worn surface.

Surface Fatigue

Surface fatigue wear is observed during repeated rolling or sliding over a wear track. The repeated loading and unloading cycles to which the material is exposed may induce the formation of surface or subsurface cracks. Eventually this will result in the loss of large material fragments due to pitting and delamination. The cracks can originate from below or at the surface. Subsurface cracks are generated at locations of maximum shear stress, increased by internal voids or inclusion. (Figure 3.12c). The initiation of cracks from the surface is caused due to oscillating compressive and tensile stresses at the area of contact and propagate into the material. Here, the cracks preferably start at 'stress raisers', such as surface inclusions.

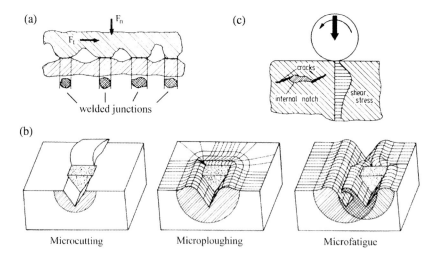

Figure 3.12:

(a) Formation and disruption of interfacial adhesive bonds; e.g. cold-welded junctions between the asperities of two metals in contact

(b) Physical interactions on the surface due to abrasion. Reproduced from [3]

(c) Crack formation and propagation due to surface fatigue

Tribochemical Reaction

The mechanism of tribochemical wear is usually discussed when materials are exposed to a corrosive environment. It results from the continual removal and new formation of chemical reaction products at the surfaces due to mechanical action. Although the knowledge of chemical or corrosive wear of polymers is even less well established than for metals, it still might play an important role in the process of surface break-down.

Most frequently the wear testing of polymers is carried out by sliding a polymer sample across a more or less rough metal surface. The predominant mechanisms are adhesion and abrasion depending on the smoothness (or roughness) of the metal. Figure 3.13 shows that there is an optimum roughness of the counterface in the dry condition, where an increase or decrease in surface roughness will generate more wear [24].

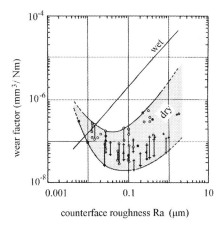

Figure 3.13: Influence of counterface roughness on wear of UHMWPE. Reproduced from [24]

It should be noted that the wear process itself can be very complex. On many occasions one of the mechanisms of wear acts in such a manner that it affects the others. Then it becomes very difficult to disentangle the complex situation and find the primary reason for wear. For example it has been discovered that hard inclusions of a first body may adhere to the softer counterbody and from there scratch the surface of the harder body. In that case abrasion is not the predominant problem, rather the adhesion of the hard inclusions [13].

3.5.2 Wear modes

The general mechanical conditions under which the system is functioning when wear occurs is termed wear mode and are listed in Figure 3.14 according to DIN 50320.

The interaction of the elements of a tribosystem may vary widely and depending on the dynamics of the system one can distinguish between sliding, rolling, impact wear, etc. The wear mode is not a steady-state condition and may transform from one to another. For example particulate debris generated by two-body abrasion might function as interfacial medium and turn the problem into a particle (= third-body) related phenomenon.

Systemic Structure	Tribological Operating Condition		Wear Mode	Acting Mechanism (single or combined)			
				Adhesion	Abrasion	Fatigue	Tribo-chem Rk
solid body – lubricant – (total separation) solid body	Sliding Rolling Wälzen Impact		–			●	○
solid body – solid body	Sliding		Sliding Wear	●	○	○	●
	Rolling Wälzen		Rolling Wear	○	○	●	○
	Oscillating		Fretting Wear	●	●	●	●
	Impact		Impact Wear	○	○	●	○
solid body – particles			Impact Erosion		●	●	○
solid body – solid body + particles	Sliding		Three Body Abrasion "Korngleit-verschleiß"	○	●	●	○
	Rolling Wälzen		Rolling Abrasion "Kornwälz-verschleiß"	○	●	●	○

Figure 3.14: Wear modes (excluding erosion) according to DIN 50320. The prevalent wear mechanisms are shown for each mode. Please note that this table is based on "general" observation. Also other combinations of wear mechanisms may apply for the specific wear mode. Adapted from [9]

3.5.3 Third Bodies

Abrasive wear may be classified as two-body or three-body abrasion according to Figure 3.15. In three-body abrasion the interfacial elements produce damage on the surfaces of the first and second bodies. In three-body abrasion only a small proportion of the abrasive particles cause wear because free rolling and sliding particles vary their angle of attack during motion (Figure 3.16). Thus, third-body wear is usually smaller than that of two-body abrasion [3].

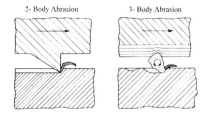

Figure 3.15: Two-body and three-body wear

The properties particle hardness, shape and size define the relative volume loss due to third bodies. In most circumstances the third bodies need to be harder than the contacting surfaces in order to cause serious damage [10, 17]. Their degree of roundness defines the angle of attack at the wearing surfaces and, thus, the occurring mechanism (microcutting re. microploughing). In addition, size is an important factor, since small particles cause proportionately less deformation than larger ones [17]. *In vivo*, the hard particles may result from X-ray contrast media (either barium sulphate or zirconium dioxide) incorporated in the acrylic bone cement. They induce abrasive third-body wear on metal and polyethylene components [33].

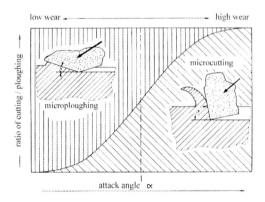

Figure 3.16: Ratio of microcutting to microploughing as a function of the attack angle of the counterbody. Reproduced from [3]

3.6 Lubrication

Wear can be reduced by lubrication. The principal idea behind lubrication is to interpose a material between two solids to minimize interaction between them. For example wetting of the surfaces reduces adhesion. The extent of fluid film formation plays an important role in the wear process of artificial joints *in vivo*. Dowson *et al.* [24] pointed out that under wet conditions the wear rate of UHMWPE decreases steadily with decreasing counterface roughness (Figure 3.13), whereas under non-lubricated (dry) conditions an optimum roughness exists as discussed previously (section 3.5.1). Hence, adequate surface finishing of the counterface (at least $R_a < 0.05$ μm) without imperfections is recommended when prosthetic materials are paired with UHMWPE [32].

The effectiveness of a lubricant film can be defined by the specific film thickness λ which is dependent on the viscosity of the lubricant, the relative velocity between first and second body, the pressure across the interface and, last but not least, on the roughness of the mating surfaces. λ can be used to estimate the occurrence of different lubrication regimes, as shown in Figure 3.17. This graph is quite similar to the better known STRIBECK curve where the coefficient of friction is plotted against the quantity $\eta \cdot u/W$, where η is the viscosity of the lubricant, u the peripheral speed of the journal bearing and W the load per unit width carried by the bearing.

In the case of boundary lubrication, the lubricant adheres chemically to one of the surfaces, and there is full contact between the solids, in contrast to hydrodynamic lubrication where a total separation of the two bodies takes place. Elastohydrodynamic (EHD) lubrication occurs when the pressure in the fluid film is sufficiently high to deform the asperities of the solid surfaces. Thus, even if the thickness of the fluid film is smaller than the heights of the asperities of first and second bodies, a total separation may be still achieved. Under realistic loads and in the presence of synovial fluid, the lubrication regime in artificial hip joints (plastic on metal) has been determined to be a mixed film [25] or boundary [26] lubrication, and solid body contact is common.

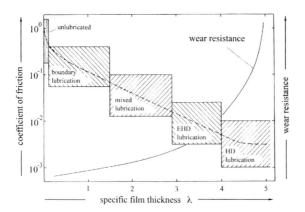

Figure 3.17: Coefficient of friction and wear resistance in rolling-sliding contact as a function of the specific lubricant film thickness. Adapted from [3]

4 System Analysis of Wear at the Tibio-Femoral Joint: A Literature Survey

4.1 Failure of Total Knee Arthroplasty

4.1.1 Reasons for Revision

There are different reasons for the primary revision of total knee arthroplasty (TKA), as reported in the Swedish register [1], whereby component loosening is still the most common indication (Figure 4.1). This type of failure can be initiated by an overload of the underlying bone inducing high stresses at the fixation interface and subsequent bone resorption. Technical errors, such as the use of highly constrained articulations which prohibit

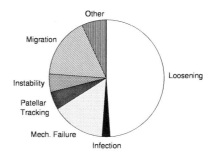

Figure 4.1: Reasons for revision during 1988-1992 as reported in the Swedish register [1]

natural rotation during flexion and extension, and the incorrect alignment of the prosthesis due to poor equipment, were common problems during the 70's and early 80's leading to this failure scenario [2]. Therefore, more anatomical shaped knee prostheses and sophisticated surgical tools for limb alignment have been developed. However, mechanical failure of TKA still continues to be a problem, although it should be recognized that loosening is no longer an issue purely related to overload at the fixation interface. Aseptic loosening due to wear and consequences of wear has emerged to be the leading cause of revision in total joint replacement [3,4].

4.1.2 Biologic Response to Wear Products

Although the bulk bio-materials are of excellent biocompatibility, it is now widely accepted that the particulate debris generated during wear is the primary cause of foreign-body induced osteolysis and subsequently aseptic loosening of the joint [5,6]. Clinically, osteolysis is recognized as diffuse cortical thinning or a focal cystic lesion in the periphery of the implant and has been reported for both total hip and knee prostheses [7-9]. The isolation of cement particles in the periprosthetic tissue around both stable and loose prostheses gave rise to doubts regarding the concept of cemented implant fixation. It was thought that the reaction of the tissue to cement debris plays a key role in the occurrence of bone loss and the term "cement disease" was created [10]. However, bone lysis was also observed with non-cemented implants, suggesting that cement is not the only reason for bone resorption around prostheses [11]. Thus, it has become obvious that the failure mechanism of aseptic loosening is not a single factor related phenomenon, but of multifactorial nature. The combination of inappropriate loading of the prostheses and the accumulation of wear debris leads to the problem of aseptic loosening [12]. While the relative contribution of each of those single factors is still under discussion, more recently, the central role of polyethylene wear debris in the production of osteolysis has been stressed [13,14]. Therefore, size, shape and morphology of polyethylene particles found in tissue around total knee implants have been well described [15-19] and their host response has been intensively investigated using in-vitro or animal models [5,20-24].

Despite the considerable amount of data in the literature, the distinct pathways of cellular response to wear particles leading to bone resorption are still unknown. One model of particle induced osteolysis has been described by Willert *et al.* [13,25] and is summarized in Figure 4.2.

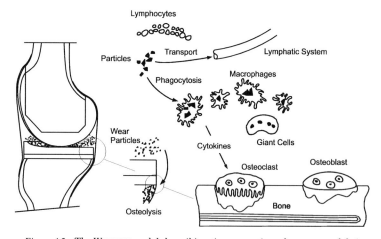

Figure 4.2: The WILLERT model describing tissue reactions due to wear debris

According to the WILLERT model, particulate debris, produced at the articulating surfaces of the prosthesis, will be phagocytosed within the joint capsule and cleared by transportation via the lymphatic system. Small amounts of foreign material can be carried away without any difficulty and an equilibrium of material wear and tissue reaction can be achieved. However, if the generation of debris overcomes the local cellular capacity to eliminate it, the particulate debris will accumulate in the capsule and subsequently infiltrate the periprosthetic tissue. The accumulated particles are responsible for the initiation of a foreign-body reaction which results in the invasion of inflammatory cells and the formation of granulomas. In particular, the interaction of smaller wear products with the monocyte macrophages causes the release of soluble inflammatory mediators which are thought to act directly on the cells responsible for bone remodeling. The resulting bone loss can be manifested as osteolysis and finally leads to loosening of the artificial joint.

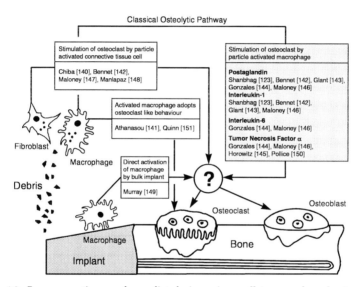

Figure 4.3: Bone resorption can be mediated via various cell types and mechanisms. The figure displays classical and alternative pathways currently under debate. Adapted from [26]

Nowadays it is recognized that the effect of bone resorption can be mediated via various mechanisms and cell types, although a consensus has developed in the literature that the classical WILLERT model is the principal osteolytic pathway. A comprehensive summery of particle induced osteolysis has been recently given by Wilson [26] and is shown in Figure 4.3. There is extensive evidence now that macrophages become active when particles are internalized. This process of phagocytosis causes the release of cytokines which stimulate the osteoclasts. As an alternative to this pathway it is possible that the macrophages adopt a osteoclast like morphology and begin bone resorption themselves when they become activated by particles. It has been also noted that cells of non-immunological lineage are capable of secreting cytokines in response to the uptake of particles. This observation suggests that the peri-implant connective tissue may contribute to the foreign body inflammatory response in addition to the dedicated immune cells. To complete the picture, it should be noted that it is also proposed that macrophages become activated by bulk biomaterials regardless of the uptake of particulates.

Further pathways of particle induced osteolysis are currently under investigation. Thus, it is hypothesized that wear particles could affect the cells of the osteoclast lineage directly, or even osteoblasts might become stimulated to produce osteoclastogenic factors by a challenge with wear particles (Hofstetter, W. Dept. of Clinical Investigation, University of Bern, Switzerland, personal communication, 1998).

4.2 Factors Influencing Polyethylene Wear in TKA

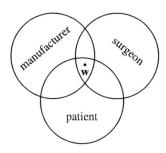

Figure 4.4: *Manufacturer, surgeon and patient control polyethylene wear*

Considerable attention has been focused on the factors which govern the wear rate of the tibial plateau since UHMWPE has proven to be the "weak link" in the TKA bearing couple. As outlined in the previous chapters, polyethylene wear is a system property and can be influenced by the manufacturing method, surgical technique, and patient characteristics (Figure 4.4). Thus, all three factors together – the manufacturer, the surgeon and the patient – define the material and mechanical factors that control UHMWPE wear. The former are associated with the material *per se* while the latter are determined by the "operating variables" (*Beanspruchungskollektiv*) of the tribosystem (see Figure 3.1).

4.2.1 Material Factors

Type of UHMWPE resin

The synthesis of the nascent UHMWPE powder (= resin) is accomplished using ethylene gas and a Ziegler-Natta-catalyst (titanium-chloride and diethyl-aluminum-chloride). Both are brought into reaction in an organic environment (typically Diesel oil). The reaction of the heterogeneous system results in UHMWPE particles ranging from 50 to 500 µm. Depending on the processing parameters different grades of UHMWPE are achieved. The powder particles vary not only in physical properties but also in morphology [35]. The latter is of specific interest, since the powder particle is composed of spherical sub-particles connected by numerous nanometer size fibrils (Figures 4.5). These are thought to be the origin of submicron size debris [36].

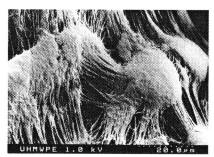 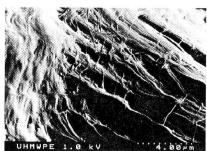

Figure 4.5: The nascent UHMWPE powder particle (usually 250 μm in diameter) is composed of subparticles connected by nanometer size fibrils.

In some grades of UHMWPE, calcium stearate is added to the nascent powder. The primary reasons for this additive are reported by [34,137,138]: calcium stearate neutralizes the electrostatic charge of the UHMWPE resin and, thus, reduces the tendency for sticking and agglomeration during handling of the material. It improves flow in ram extrusion and prevents corrosion of the molding equipment. In addition it hinders yellowing of the resin during various fabrication steps. From the standpoint of polyethylene wear in total joints the incorporated additive is of concern because calcium stearate is suspected to act as third-body [37]. Further there is evidence that calcium stearate may interfere with optimal consolidation of UHMWPE and increase oxidation at the boundaries between resin particles [139].

The majority of UHMWPE used in total joints has originated from three types of resin, GUR 1120, GUR 415, and 1900 resin [137]. Today, the 1900 resin which was manufactured by HIMONT (Wilmington, Delaware) is no longer on the market. Since several years, the only available UHMWPE grades for medical applications have been processed by HOECHST (either Oberhausen, Germany or Houston, Texas). Still, the quality might have differed from product to product since the synthesis is not alike between Oberhausen and Houston (Dr. Haftka, Ruhrchemie, personal communication). TICONA, a recent business spun off by HOECHST, currently produces four grades of UHMWPE certified to ISO 5834 part I and ASTM F648: GUR 1150 (previously 415),

GUR 1120 (previously often called 412), and with the general move away from calcium stearate, GUR 1020 and GUR 1050*.

Table 4.1: *Physical properties of the three main grades of UHMWPE resin used in orthopedic devices (data from [34])*

Type of resin	Molecular Weight [g/mol]	Crystallinity [%]	Melting Point [°C]
1900†	$2 - 4 \times 10^6$	75	145
GUR 1120	4×10^6	60	143
GUR 415	6×10^6	58	143

Among the different grades of resin, there are two ranges of molecular weight available, 2 to 4 million and 4 to 6 million. While the former shows higher molecular chain mobility and higher toughness, the latter provides better abrasion resistance. The physical properties listed in Table 4.1 seem to have notable influence on the wear of UHMWPE. For instance, Furman *et al.* [27] found a significant effect of the type of resin on the oxidative behaviour. Oxidation of the polymer surface causes embrittlement and increases the possibility of crack initiation (and, hence wear). The quality of the processed bar stocks is also determined by the type of resin. Mayor *et al.* [28,29] reported the absence of fusion defects in retrieved tibial inserts that had been fabricated from GUR 415 while defects were found in inserts fabricated from 1900 resin and GUR 412. Weightman and Light [30] found a lower wear rate of Hi-Fax 1900 as compared to RCH 1000 (= sheet molded GUR 1120) while Walker *et al.* [31] demonstrated inferior wear characteristics of RCH 1000 versus GUR 415. As shown in Table 4.1, the resin types differ mainly in molecular weight and crystallinity: the higher the molecular weight, the lower the crystallinity. According to Eyrer [32], increasing crystallinity (below a level of approx. 70%) leads to increased resistance to crack initiation, crack propagation, and oxidation, while based on the results of Rose *et al.* [33], higher molecular weight induces a better wear resistance in total joints.

Although many authors tried to stress the intrinsic wear properties of the specific type of resin, it should be noted that most of them were not investigated independently from the

---

* GUR number system: *first digit* = 1, resin from Hoechst Germany, if = 4, Hoechst Texas; *second digit* = 0, no calcium stearate, if = 1, contains calcium stearate; *third digit* = 2, molecular weight = 2 to 4 million, if = 5, molecular weight = 4 to 6 million.

† 1900 resin was 2 to 4 million molecular weight in 1970s. It was increased to 4 to 6 million in 1980s.

applied processing technique. For example, GUR 415 is available as ram extruded product only, whereas RCH 1000 is a sheet molded device.

Consolidation of UHMWPE

The principal variables in consolidating UHMWPE resin are time, temperature, pressure, and the method of consolidation [137]. Currently there are three methods of consolidating UHMWPE resin. The first method is direct molding, in which the powder is placed into a mold that compresses it into the final shape of the medical device. The second method involves the molding of large sheets of UHMWPE which can be as thick as 20 cm. From these sheets the implant is then machined. The third method is also a two-stage process. A cylindrical bar stock is produced by ram extrusion which is later machined for the final product. Due to the high viscosity of UHMWPE, all methods involve simultaneous heating and pressurizing of the nascent powder. The fundamental difference between compression molding and ram extrusion is the way of processing. Ram extrusion is continuous while compression molding is static. This implies that the variables associated with ram extrusion are more complex (e.g. adjustment of the pressure profile of this open-ended process) and therefore more difficult to control [137]. Table 4.2 summarizes the history of both the main resins and consolidation methods. Today, there are four companies which sell consolidated UHMWPE to the medical market: PERPLAS MEDICAL (UK), PPD MEDITECH (USA), POLY HI SOLIDUR (USA), and WEST LAKE PLASTICS (USA).

The unit costs of compression molded components are much higher than those of ram extruded parts and, therefore, the latter have dominated the U.S. market since the late 1980s[*]. Although the mechanical characteristics of both products tend to be equal [38], the wear properties of ram extruded UHMWPE seem to be inferior. Huber *et al.* [39,40] found a clearly higher wear rate of ram extruded versus molded polyethylene conducting pin-on-disc wear tests. These results appear to be reflected *in vivo*: Bankston *et al.* [41] determined radiographically half the wear rate of direct compression molded hip cups as compared to ram extruded components. These unsatisfactory wear characteristics may be explained by defects occurring in the bulk polyethylene. It is known that ram extrusion is susceptible to the creation of voids in the bar stock because of the intermittent pressurization changes during fabrication [37]. Tanner *et al.* [42] found intergranular separation, fusion defects, and large subsurface cracks exclusively in ram extruded

---

[*] Just recently, compression molded GUR 1020 and 1050 are beginning to supersede ram-extruded GUR 4150.

components, and not in direct molded knee components of the same type (Ortholoc II). Interestingly, the former retrievals showed a significantly larger delamination rate, which highlights the importance of the fusion process of the powder particles (Figure 4.6).

Direct compression molded parts often show large variations in crystalline morphology which also may affect the wear characteristics. The crystalline regions of UHMWPE usually consist of lamallae and do not organize further to form spherulites [34]. In compression molded implants, however, spherulitic structures may be scattered within amorphous zones as documented by the author (Figure 4.7) and the report of Jasty *et al.* [43]. In addition, the cooling process after compression molding can produce tensile residual stresses within the bulk [44].

Table 4.2: *Dominant UHMWPE resins and principle consolidation methods. From [137]*

Time Frame	Resin	Consolidation	Trade Name
1960s to 1990s	GUR 1120	sheet molding	Chirulen and RCH 1000
1972 to early 1980s	1900 resin	sheet molding	Hercules 1900
Early 1980s to 1995	Himont 1900	direct molding	Himont/ Hifax 1900
Early 1980s to 1990s	GUR 415	ram extrusion	GUR 4150 extruded bar
Mid- 1990s	GUR 1020 and 1050	sheet molding	GUR 1020 (1050) compression molded sheet

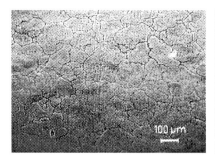

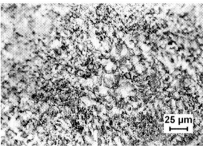

Figure 4.6: Transmission light micrograph of ram extruded polyethylene under phase contrast; note that the grain boundaries of the nascent powder are still present.

Figure 4.7: Transmission light micrograph of compression molded polyethylene using interference contrast; note the organization of spheruletic structures.

Surface finish

Machining marks on the polyethylene surface (Figure 4.8) are suspected to be a major source of early particulate debris during running-in [45]. As has been shown by Bristol *et al.* [46], local non-uniformities of contact stress are much more pronounced for machined versus molded components and, hence, the former may lead to a higher wear production. To address this problem, there have been attempts made to smooth the surface of ram extruded, machined implants by applying a heat pressing method. The clinical results of this method were fatal. Instead of a reduced wear generation, complete failure of those inserts were common [47,48]. Remaining residual stresses between the substrate and the heated layer lead to early fatigue and catastrophic failure of the polyethylene insert.

Figure 4.8: Machining marks on a "off-the-shelf" tibial liner

Sterilization and packaging

According to Lewis [37], the current method of choice of sterilizing UHMWPE is gamma irradiation in air at 20 to 50 kGy (1 Gray = 100 Rad) using a ^{60}Co source. The effects of this sterilization technique on the mechanical properties of the polyethylene have been quite well understood for some time [49-51], however, its effects on wear characteristics are still under debate [52-54]. Several models aiming to explain the clinically observed degradation process exist [55-57]. The gamma radiation directly attacks the UHMWPE chains causing the formation of radicals by breakage of the carbon-hydrogen or carbon-carbon bonds. Thus, either cross-linking or chain scission occurs. The latter is agreed to be the dominant mechanism, especially under an oxidative environment. This oxidative degradation, which is assumed to be a combined chemical and mechanical effect [58,59], results in higher density and crystallinity with time [52,60]. Since the latter reflects an increase in the elastic modulus, stresses occurring inside the implant will rise as well.

Increases in stresses are associated with wear damage. Applying a pin-on-disc test, Fisher et al. [61] was able to demonstrate a nearly 5-fold increase in wear of irradiated, artificially aged polyethylene versus non-irradiated polyethylene. However, there is growing concern that the applied aging protocols for accelerated laboratory tests may not reflect the in-vivo situation. Interestingly, the oxidation peaks of retrieved implants do not occur at the surface (as due to artificial aging) but beneath the surface [62]. After cross-sectioning, those oxidized zones in implants appear as "white bands". They represent areas of embrittlement. Applying tensile tests to thin cross-section of acetabular liners, the components failed first in the white band region with a brittle fracture [54]. Sutuala and coworkers [54] were able to demonstrate that cracking and delamination of clinically retrieved components were highly dependent on the presence of the subsurface white band, whereas other modes of wear did not show any significant correlation. This observation is consistent with the observation of Bell et al. [63] using pin-on-plate wear tests: delamination only occurs in UHMWPE specimens where a subsurface oxidation maximum is developed.

In order to overcome these problems of material degradation, it is suggested to either gamma sterilize and pack the UHMWPE components in an inert atmosphere (e.g. nitrogen, argon) [64], or use ethylene oxide treatment [65], another "cold" sterilization process for medical devices. While there are divergent opinions about the former treatment (*in vivo*, the degradation process continues due to the presence of oxygen in the synovial fluid [66]), components treated with ethylene oxide did not show the presence of white bands at their time of retrieval [54]. Although this obviously demonstrates that the damage is reduced, it is not yet clear whether it will provide a clinically meaningful improvement over the life of the implant [34].

Countersurface

The surface properties of the opposing component have substantial impact on the wear of the polyethylene bearing couple. Therefore, these properties and their relation to polyethylene wear represent an issue associated with voluminous research [67-83]. In particular, counterface roughness has quite recently gained attention again [76-78]. This renewed interest is a direct result of new measurement utilities which allow a fuller description of the topography (see section 3.3.1). Theoretical approaches towards wear particle generation are nowadays capable to predict reasonable numbers [76,78]. Experimentally, Dowson et al. [79] noted already that there is an optimum roughness which produces a minimum rate of wear of UHMWPE under dry sliding conditions. In the presence of water, however, the wear rate of the polymer decreases steadily with counterface roughness (Figure 3.13). Fisher et al. [80] found a similar variation in wear rate with counterface roughness using bovine serum as the lubricant. It should be noted that even the grinding direction [81] or single scratches [82] influence the wear rate and morphology [83] of the generated debris to a considerable degree.

As reviewed by Goodman and Lidgren [67], the current material of choice of the femoral condyles is cobalt-chromium alloy. It is assumed to cause less polyethylene wear *in vivo* since the surface is less susceptible to third-body particles [68,69,73]. However, clinical studies are inconclusive in this respect. Kraay *et al.* [70], for instance, have not observed significant problems associated with titanium as a bearing material in total knee replacement. In addition to metal components, ceramic condyles are used. There are several wear tests which emphasize the improved tribological behaviour of zirconia versus metal articular surfaces [31,71-73]. The superior characteristics, namely a lower coefficient of friction and a reduced rate of wear, are attributed to a better wettability, minimized adhesion, and the resistance of the hard ceramic surface against roughening.

Lubrication Properties

In laboratory tests higher wear rates have been observed for distilled water compared with serum, attributed mainly to the occurrence of polyethylene transfer films with the former lubricant [84,85]. For example, the reduction of serum protein in the lubricant to as low as 20% caused marked differences in the type and amount of polyethylene wear generated in a hip simulator [86]. A possible reason for this behaviour is the attachment of serum proteins to the metallic surface which prevent the build-up of transfer films and improve the boundary lubrication characteristics [87].

These lubrication differences may be also reflected *in vivo*, since the viscosity and composition of synovial fluid around artificial joints varies considerably and is often different from "normal" due to the capsulectomy involved in total joint replacement [31,88]. To study the effects of these differences on the tribology, McKellop *et al.* [89] assessed the lubricating properties of joint fluids using a pin-on-disc machine. Unexpectedly, the results indicated that the lubrication of polyethylene/ cobalt-chromium bearings was not markedly affected over the range of pH, viscosity or protein content of the measured samples. Thus, it was concluded that (a) proteins do not act as boundary lubricants or (b) even the lowest concentrations of occurring proteins are sufficient to saturate the surfaces with boundary lubricant.

4.2.2 Mechanical Factors

Contact load

The contact pressure dependence of the wear rate of UHMWPE has been described by several authors. Rose *et al.* [33] as well as Rostocker and Galante [90] proposed an exponential relationship between applied load and wear behaviour of polyethylene which was experimentally confirmed for the application of total knee replacement by Walker *et al.* [91] and Treharne *et al.* [92]. In a later study by Fusaro [93] it was outlined that the UHMWPE wear rate is only dependent on contact stresses above 13 MPa while below this level there is no influence.

The applied load can be static or cyclic. Walker *et al.* [31] found that the weight loss for static and cyclic loading is comparable up to 2 million cycles using a reciprocating pin-on-flat device. However, the wear for static loading exceeded that for cyclic loading beyond 2 million cycles. While the aforementioned study focused on surface wear, Pruitt *et al.* [94] demonstrated that cyclic loading can result in fatigue cracks within the bulk polyethylene. Interestingly, not only tensile-compressive load cycles lead to crack initiation but also compressive-compressive cycles. For both conditions the initiated fatigue cracks propagate to a greater extent as the load amplitude is increased.

Contact kinematics

There is common agreement that the wear volume of polyethylene increases linearly with sliding distance [93,95]. Thereby, several authors have noted that the specific contact kinematics (such as uni- and bi-directional motion) may play a crucial role for the generation of polyethylene debris [96,97]. This is an interesting observation because the femoral condyles move in a complex combination of rolling, sliding and spin during knee flexion and, thus, there is uncertainty about the relative amount of damage produced. Experiments comparing the effects of rolling and sliding concluded that sliding causes higher damage to the polymer than rolling [98]. In addition, it was highlighted by many authors that crossing motion trajectories* during sliding movement accelerate the wear of polyethylene [99-101]. For the knee joint, these crossing contact paths occur when the antero-posterior translation is superimposed with internal/ external rotation. Walker *et al.* [31] demonstrated that the wear rate increases significantly under these conditions.

---

* Bi-directional oscillation and phase difference generates crossing motion trajectories at the contacting surfaces [100].

Patient characteristics
The aforementioned factors – contact load and kinematics – are in part driven by patient characteristics and should be reflected in clinical wear studies. Although many of these studies demonstrated a correlation between damage and time in situ of the implant [28,47,102-105], most of them failed to show any further correlation with other clinical data (i.e. weight, age, sex, diagnosis, etc.). This observation at first seems inconsistent with the mechanics at the knee and subsequent wear. For example, a heavy patient should generate higher contact loads at the articulation than a skinny person and, thus, produce more damage at the tibial plateau. However, it has been shown that wear is a function of (total) sliding distance in the first place. This can be better described by patient activity – which has been shown to vary by as much as 45-fold [106] – than by data of the clinical record. In fact, clinical data do not correlate with the activity of the patient as described by a study of Seedhom and Wallbridge [107].

Gait adaptations, common to patients following TKA [108], may influence the dynamics of the artificial knee joint [109]. Hilding *et al.* [110] reported that patients who walked with a higher peak flexion moment showed increased tibial component migration which put them at risk for aseptic loosening. Andriacchi *et al.* [108] found distinct abnormalities in the gait of total knee patients. These abnormalities were not only influenced by the patient himself but also by TKA design: patients who were treated with the least constraining design were the only group which had a normal range of motion during stair climbing.

Implant design
The amount of constraint in the prosthesis can markedly affect the contact mechanics of the joint. Less constraining designs are usually posterior cruciate retaining (CR) implants while posterior stabilizing (PS) prostheses are more constraining. Using fluoroscopic analysis during steady state treadmill gait, Banks *et al.* [111] found significant differences in the contact kinematics between CR and PS designs, while Taranowski *et al.* [112] outlined that only CR knees may achieve displacement patterns similar to unoperated knees. It has been suggested that the antero-posterior movement of the femur on the tibia influences function during daily activities [113] and, thus, may explain the gait abnormalities observed in PS designs.

It is generally accepted that higher conformity of the tibial plateau reduces stresses inside the bulk polyethylene due to the larger contact area [114,115]. On the other hand, freedom of motion is limited, which increases the constraint forces on the tibial plateau [116]. Therefore, in order to minimize surface damage, Bartel *et al.*

[117] suggested that conformity should be provided rather in the frontal plane than in the sagittal plane (Figure 4.9a): a single radius of curvature in the frontal plane will provide a large contact area and minimize stresses. However, the medio-lateral lever-arm can be insufficient to prevent lateral joint opening [118] (Figure 4.9). This lever-arm is enlarged with a cylinder-on-flat design (thereby keeping conformity in the frontal plane). Although lateral joint opening is now less likely, the risk of edge-loading is increased [119] (Figure 4.9b).

The impact of the above design issues on tibial component wear and clinical outcome is still unclear and subject of much debate [120-122]. Low conformity (in the sagittal plane) was found to produce more wear and less clinical success in the studies of Collier *et al.* [123] and Feng *et al.* [124]. This was not observed by Blunn *et al.* [125,126] and Ritter *et al.* [127]. Ritter even promoted low conformity as a primary reason of success: the cylinder-on-flat design allows freedom of rotation and translation which reduces constraint forces and, thus, surface damage.

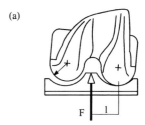

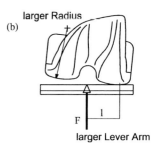

Figure 4.9: AP-view of the femoral curvature in total knee arthroplasty. The larger radius of design (a) will provide a larger lever-arm to balance the adduction moment and, thus, increases the stability against lateral joint opening. While both designs allow maximum conformity in the frontal plane, the risk of edge-loading is increased in design (b). Adapted from [118]

Finally, the thickness of the tibial polyethylene component is also an important design feature. Thickness of polyethylene inserts has become an issue because it was shown that metal backing inhibits cold flow of the polymer and improves the load transfer to the bone [128]. In addition, metal backing allows anchoring of the prosthesis without usage of bone cement (due to direct bone ingrowth). In order to avoid excessive tibial resection the use of thinner tibial components became necessary. However – as has been shown by finite element studies – the thickness of the UHMWPE component should not be decreased below 6 mm, where the stresses increase exponentially [117,129]. This increase in stress of too thin tibial plateaus leads to severe damage as has been documented by several retrieval studies [123,130].

Surgical technique

In a recent study of Schmalzried *et al.* [131], analyzing the *in vivo* wear of over 1000 artificial hips from radiographs, the variable with the greatest effect on wear was the surgeon. The eleven surgeons participating in this multicenter study explained 70% of wear variability (the second highest variable explained only 35%). This illustrates the critical role of surgical technique. Since the biomechanical reconstruction of the knee is even more demanding than the hip, surgical technique of TKA has presumably similar if not more influence on wear. Malalignment and improper soft tissue balance are commonly reported and are known to cause elevated and eccentric loads between femur and tibia [132]. This increases the contact stresses tremendously often resulting in disastrous wear [123,133,134].

Based on cadaver studies, it was recently argued that contemporary alignment techniques in TKA do not produce accurate rotational alignment [135]. This seems to be mirrored by a retrieval study by Wasielewski *et al.* [136] who analyzed the wear patterns of unconstrained tibial inserts and found that most of these inserts displayed asymmetric wear between the medial and lateral compartment of the plateau (wear was posteriorly located in the medial compartment and more anterior in the lateral compartment). In addition to joint line position, knee alignment and ligament balance, scratches on the articulation or generation of third body abrasive particles from extraarticular sources are under the surgeon's control at the time of implantation.

In summary, there are multiple and interrelated factors that can affect the wear of tibial UHMWPE components. The complexity of the system is once more illustrated in Figure 4.10. It demonstrates the difficulty in controlling the variables in order to achieve meaningful results from retrieval analysis and wear test studies. Thus, the understanding of the complexity plays an important role in the identification of the prevalent wear mode and wear mechanism of the tribosystem. Those are still not understood completely, despite the enormous effort in the literature. To the author's opinion, this might be related to an inappropriate description of the system (e.g. dynamics of the prosthesis) on the one hand and, on the other hand, the incomplete analysis of the *early* wear regime (prior to failure). The cognition of the latter may provide the key to enhance the wear properties of the tibial polyethylene component.

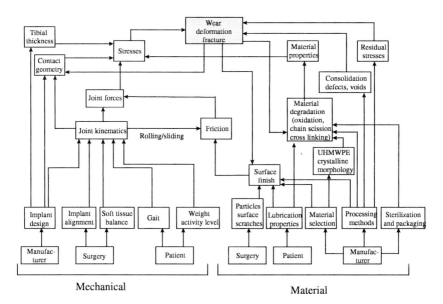

Figure 4.10: Wear at the artificial knee joint applying the system approach

5 Contact Mechanics at the Tibial Surface

5.1 Introduction

5.1.1 The Role of Friction in the Analysis of Joint Dynamics

In section 2.3.2 the models of Morrison [1], Seireg and Arvikar [2], and Schipplein [3] were recapitulated. While these knee models have been employed mainly to evaluate the bone-to-bone contact forces, various studies of different complexity describe the articular kinematics and kinetics in more detail. Andriacchi et al., Wismans et al., Essinger et al. and Blankevoort et al. [4-7] determined the relative movement of the femur on the tibia using the direct dynamics approach where the forces were input and the motions output. Other investigators used defined motion patterns of the femur on the tibia to study the effect of relative movement of the articular surfaces on the produced forces [8,9]. All the models above were applied to the normal knee and none of the models took surface friction into account.

The coefficient of friction of cartilage on cartilage is extremely low (0.0028...0.0054 [10,11]) and, thus, forces due to friction can certainly be neglected in the normal knee. In contrast, reported values of friction in artificial joints range from 0.04 to 0.2 [12-16] and can even rise to 0.35 if "lumpy" polymer transfer occurs [12]. The increase in friction has important implications for the contact mechanics of a total knee arthroplasty. Friction introduces shear forces at the articulating surfaces, potentially damaging the polyethylene [17]. These tangential shear forces can be generated not only during sliding but also during rolling motion as will be shown in the next section.

5.1.2 Tractive Forces during Rolling Movement

Pure rolling occurs when the relative velocities v_1 and v_2 of two rigid bodies at the point of contact are equal (Figure 5.1). If they are unequal, the rolling motion is accompanied by sliding. The term pure rolling, however, might be misleading since the absence of apparent sliding does not exclude the transmission of a tangential force $\bar{F}_t$ of magnitude less than limiting friction [18].

To exemplify such conditions of rolling movement, one might consider the wheels of a car which undergo driving and braking and as a result tangential load transfer during starting and stopping procedures on the road. Johnson [19], therefore, suggested to use the terms free rolling and tractive rolling to describe motions where the tangential surface loads are zero and non-zero respectively.

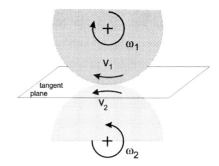

Figure 5.1: *Rolling is defined as a relative angular velocity between the two bodies about an axis lying in the tangent plane.*

The tractive force vector $\bar{F}_t$ generated during rolling motion can be best explained mechanically when the rolling body is of irregular shape (Figure 5.2). To initiate rolling movement of this body a force

$$F_t = F_n \cdot \tan \Phi \qquad (5.1)$$

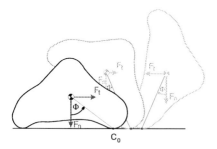

Figure 5.2: *Rolling friction force for an irregular object*

(inertia neglected) is needed. $\bar{F}_n$ is the normal load and Φ the angle between the vertical and the line joining the center of gravity of the body and the instantaneous center of rotation C_0. If the coefficient of friction between the object and the ground is less than $\tan\Phi$, the tangential force $\bar{F}_t$ will produce sliding rather than rolling. If friction is adequate, a tractive force equal to $\bar{F}_t$ will maintain equilibrium at the contact C_0. As rolling continues, Φ changes, even taking on negative values. Hence, the tractive force to maintain rolling at a constant velocity may change direction along the surface, although the direction of movement of the body may remain the same. This illustrates that, in contrast to sliding, where the vector of the sliding force always hinders movement, the vector of the tractive force is not directly dependent upon the direction of movement but rather maintains rolling equilibrium (until limiting friction is exceeded).

Commonly, the tractive force generated during tractive rolling can be substantial since it is governed by a friction coefficient μ_{max} exceeding the dynamic coefficient μ_d applicable during pure sliding. There is a paucity in the literature regarding the behaviour of the coefficient of friction versus different degrees of rolling and sliding in the metal-polyethylene articulation. From the car industry, however, it is known that the highest friction of tires does not apply during pure sliding, but at a slide-roll ratio of 30 -70% depending on the road conditions (dry, wet, snowy, etc.) [20]. The latter finding drove the development of anti-lock braking systems (ABS) and emphasizes the importance of this physical effect for the transfer of friction. In this study, it was assumed that μ_{max} is identical with the static coefficient of friction μ_s (which is twice the dynamic coefficient of friction μ_d for a metal-polyethylene articulation [21]). This assumption is based on the idealistic theory of pure rolling[*]: the instantaneous relative velocity at the contact is zero. Hence, the initiation of sliding motion is governed by the static coefficient μ_s.

5.1.3 Purpose

The purpose of this study was to determine normal and tractive forces in the sagittal plane during rolling motion at the artificial tibio-femoral articulation. Conditions that can increase (or decrease) the tractive forces at the polyethylene component, in particular, the influence of the coefficient of friction, tibial conformity, patella position, and gait mechanics were studied. Finally, in a second model, the combined effect of the calculated normal and tractive forces on the stresses in the polyethylene component was analyzed.

5.2 Materials and Methods

5.2.1 General Description

The inverse dynamics approach was used to calculate the normal $\bar{F}_n$ and the tractive forces $\bar{F}_t$ at the knee from kinematic and kinetic measures taken during the stance phase of gait of patients following total knee arthroplasty (TKA). The model of the artificial knee (Figure 5.3) was based upon previously described models of the natural knee [1,3] with the following modifications:

---

[*] It should be noted that *pure* rolling is an ideal condition and not applicable to technical systems. Typically, there is micro-slip next to stiction in the contact zone due to the compliance of the elastic bodies.

- the kinematic linkage (pure rolling or sliding) between the femur and the tibia was determined by the static and dynamic coefficient of friction and the angle of knee flexion;

- the lines of action for each muscle group were approximated by force vectors that change direction as a function of knee flexion;

- gait kinetics common to patients following TKA [22] were used.

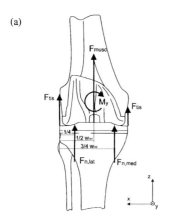
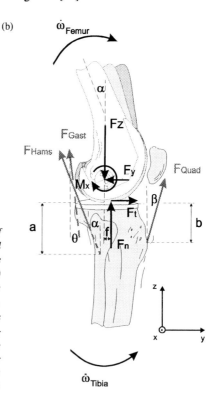

Figure 5.3: Frontal (a) and sagittal (b) view of the proximal portion of the tibia and the distal part of the femur with attached total knee prosthesis. Muscle groups (Quad, Hams, Gast) and soft tissue structures (Tis) crossing the joint are represented by force vectors. Here, the external forces ($\vec{F}_z$, $\vec{F}_y$) and moments ($\vec{M}_x$, $\vec{M}_y$) are shown acting about the center of curvature of the femoral condyles ("negatives Schnittufer"). The patellar ligament angle, β, changes with knee flexion angle, α. For a precise description of all symbols see appendix 5-I.

The model followed a statically determinate approach to balance external forces and moments at the knee joint with internal forces and moments.

$$\vec{F}_{ext} - \vec{F}_{int} = 0; \qquad (5.2)$$

$$\vec{M}_{ext} - \vec{M}_{int} = 0; \qquad (5.3)$$

While the external forces and moments were generated by foot-floor contact, gravitational and inertial forces, the internal forces and moments were generated by muscles ($\vec{F}_M$), soft tissue ($\vec{F}_{tis}$) and contact forces ($\vec{F}_C$) at the artificial articulation. The vector equation for internal force and moment equilibrium was given by:

$$\vec{F}_{int} = \sum_{i=1}^{n} (\vec{F}_M)_i + \vec{F}_{Tis} + \vec{F}_C ; \tag{5.4}$$

$$\vec{M}_{int} = \sum_{i=1}^{n} \left(\vec{r}_i \times (\vec{F}_M)_i \right) + \vec{r}_{Tis} \times \vec{F}_{Tis} + \vec{f} \times \vec{F}_C ; \tag{5.5}$$

where i was the number of active muscles per time step, $\vec{r}$ the position vector of the force producing structure to the tibio-femoral contact point, and $\vec{f}$ a rolling lever which will be explained in section 5.2.2. The medio-lateral force component and the internal-external moment were neglected in the vector equation, since these load components do not contribute to the contact force vector (consisting of $\vec{F}_n$ and $\vec{F}_t$) in the sagittal plane. A complete detailed mathematical description of the model is given in appendix 5-I.

5.2.2 Kinematic Approximation

The movement of the tibio-femoral contact point was derived from cadaver studies of knees with the anterior cruciate ligament and menisci removed [23]. During rolling, the contact point between femur and tibia moves antero-posteriorly as a function of the curvature of the articulating surfaces and the flexion angle. The curvatures of the two femoral condyles were modeled as cylinders (55 mm radius) sitting on a polyethylene tibial surface inclined to represent the anatomical posterior slope of the tibia. The influence of conformity, defined as the ratio between the sagittal radius of the femoral and tibial components, was investigated by systematically altering the radius of the tibial plateau between 110 mm and flat (conformity ratio 0 ... 0.5).

Pure rolling between the two rigid body segments was simulated between -10° and 18° of knee flexion [23]. In addition, pure rolling was limited to the condition where the ratio F_t/F_n was less than the static coefficient of friction μ_s. The maximum static coefficient of friction used in the model was twice the dynamic coefficient μ_d [21]. Pure sliding with $\mu_d = 0.1$ [12,13] occurred beyond 18° (or anytime F_t/F_n was greater than μ_s).

During femoral rollback the medio-lateral contact of the two femoral condyles on each side of the tibial plateau remained fixed at 25% of the tibial width (measured from the knee joint center). In this study the width of the tibial polyethylene liner was set to 80mm (Figure 5.3a). The antero-posterior contact path was derived from a retrieval study of the Miller/Galante implant simulated in this study [24,25], while similar data have been published by Wasilewski et al. [26] and Blunn et al. [27]. The contact point at full extension was 20mm posterior from the anterior lip of the insert. The maximum femoral rollback was set to 17mm. Rolling resistance was included by means of a rolling lever f of the joint reaction force about the center of instantaneous motion (Figure 5.3b). The rolling lever was defined such that it resisted rolling movement. The calculation of this lever was based on the material constants of UHMWPE and Cobalt-Chromium (see appendix 5-II).

5.2.3 Approximation of Ligament and Muscle Force Vectors

The simulation of the retained posterior cruciate ligament following TKA was based on a cadaver study by Mahoney et al. [28]. He found that significant contribution of the PCL in total knee stabilization did not occur earlier than 30° of knee flexion. Thus, it was assumed that the PCL did not tighten during the range of knee flexion where tractive rolling took place. The anterior cruciate ligament was sacrificed at the time of surgery and therefore was not included in this analysis. Lateral or medial soft tissue tension [3] prevented lift-off of the joint in the frontal plane (Figure 5.3a). The vectors of medial and lateral collateral ligaments, as well as tensor fasciae latea were assumed to act through the point of contact between femur and tibia in the sagittal plane.

The assumptions made with respect to muscle grouping and vector modeling were based on studies by Morrison [29] and Schipplein et al. [3]. The "quadriceps group vector" $\bar{F}_{Quad}$ included the rectus femoris and vasti muscles. The "hamstrings group" $\bar{F}_{Hams}$ formed a single vector consisting of the semimembranosus, semitendinosus, biceps femoris and gracilis muscles. The medial and lateral heads of the gastrocnemius muscles together with the plantaris muscles formed the "gastrocnemius group" $\bar{F}_{Gast}$. Using a straight-line assumption, the action lines of these muscle groups were approximated by force vectors acting on the tibia. The lines of action of the muscle groups were dependent on the knee flexion angle.

Muscle attachments and lines of action were derived from a study by Draganich [30]. The quadriceps vector followed the line of action of the patellar ligament during knee flexion and inserted 25 mm below the tibial plateau (Figure 5.3b). The hamstring vector followed the axis of the femur (and thus the knee flexion angle) at all time and inserted as a single point 40 mm below the tibial plateau. The medial and lateral head of the gastrocnemius originated proximally with respect to the medial and lateral femoral condyles. Both heads wound around the medial and lateral condyles respectively and converged distally into the calcaneus tendon. Due to femoral rollback, the angle of $\bar{F}_{Gast}$ varied from -3° to +9°, with respect to the tibial axis during knee flexion.

Patello-femoral mechanics were defined in terms of angular change of the patellar ligament. The directional change (β) of $\bar{F}_{Quad}$ followed the angular change of the patellar ligament with flexion. The angle, β, was linearly dependent [31,32] on the flexion angle, α, as illustrated in Figure 5.3c. The initial orientation of $\bar{F}_{Quad}$ at full extension (α = 0°) was taken to be 22° and 30° to represent a patella of nominal thickness, and the situation where the nominal thickness was increased by approximately 6 mm, respectively.

Using the muscle groupings described above only one agonist group was active at a time (based on EMG), which allowed a statically determinate handling of the equilibrium. The influence of antagonistic muscle activity was evaluated parametrically using the approach described by Schipplein and Andriacchi [3]. Antagonistic muscle activity was simulated based on a proportion of the external flexion-extension moment $\bar{M}_x$ needed to balance the net moment at the knee. EMG measurements were used to determine on-off activity of muscle groups. The effects of antagonistic muscle work were investigated by simulating antagonistic levels of 0, 10, 30, and 45% of the moment created by the agonists.

$$|M_x| = \left| \left(F_y \cdot r_z - F_z \cdot r_y \right)_{Agonist} \right| - \left| \left[(0, \cdots, 45)\% \cdot M_x \right] \left(F_y \cdot r_z - F_z \cdot r_y \right)_{Antag.} \right| \quad (5.6)$$

External abduction or adduction moments were balanced about the lateral or medial joint contact, respectively. The muscle forces (calculated in the sagittal plane) plus the ground reaction force (resolved along the tibial axis) resisted the external moments in the frontal plane, thereby assuming that the sum of the muscle forces, which was determined in the sagittal plane, acts through the center of knee joint (Figure 5.3a). If the abduction or adduction moment $\bar{M}_y$ was not balanced by the summed muscle and axial forces, equilibrium was maintained by medial or lateral soft tissue tension (see appendix 5-I).

5.2.4 Model Input

In addition to the knee flexion angle and EMG recordings, external moments and forces about the knee were input to the model. Data were obtained from patients following total knee replacement during stance phase of gait while level walking (raw data see appendix 5-III). The selection of characteristic subjects was based on a previous study of total knee patients [22].

The methodology used for the measurement of external motion, forces, and moments can be found in detail in the appendix 5-IV. Briefly, the lower extremities (foot, shank and thigh) were idealized as rigid-body linkage with fixed axes of rotation at the joints. The ground reaction forces, limb segment masses, and inertia were used to calculate external joint moments and forces. The set-up for the measurements included a force platform and an optoelectronic system for analysis of segment motion.

Two characteristic* gait patterns (Figure 5.4) were analyzed. The first pattern was characterized by a normal net flexion/ extension moment *("Normal")* while the second had a distinct reduction in the external flexion moment *("Quadriceps Avoidance")*. The external forces obtained in the sagittal plane already included the effect of acceleration of the lower limb segments, and the external moments took the inertial moments into account. Both the external medio-lateral force component and the internal-external moment were neglected. The output of the model, $\bar{F}_n$ and $\bar{F}_t$, was the sum of the respective forces acting on the medial and lateral plateau.

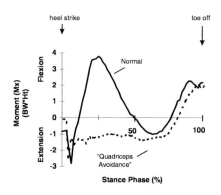

Figure 5.4: The two patterns of flexion/ extension moment $\bar{M}_x$ used for input to the model. The abnormal "quadriceps avoidance" pattern is common to patients following total knee replacement.

5.2.5 Stress Analysis

The method of finite elements was applied to evaluate stresses in the UHMWPE component using the commercially available code ADINA (Automatic Dynamic Incremental Non-linear Analysis), version 7.0. ADINA supports non-linear applications and uses a contact algorithm without the use of so-called "gap" elements. Instead, the algorithm criterion is such that the degree of mesh overlap for the most current geometries is determined and then is eliminated at each equilibrium iteration [33]. More details of the contact algorithm are in appendix 5-V.

---

* based on [22]

The two-dimensional model adopted for this study was originated and validated by Beard [34] and utilized for the analysis of residual stresses in polyethylene due to normal loading [35]. Briefly, this model consists of a femoral, elastic component which is formed as a CoCr sphere, and a tibial UHMWPE component approximated as plastic-multilinear material (Figure 5.5). The elastic modulus, Poisson's ratio and yield strength were obtained from experimental results of DeHeer [36] using ASTM D695 compression testing (Table 5.1). The stress-strain curve found by DeHeer [37] was modeled as multilinear (Figure 5.6). A von Mises yield criterion was included. Once plastified, the occurrence of residual stresses generated isotropic hardening of UHMWPE [34].

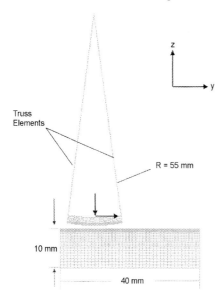

Figure 5.5: *The two dimensional finite element model adopted from Beard [34]. The bottom surface of the tibial component was constrained in all directions to prevent rigid body motion.*

Table 5.1: *Material Properties of the tibial component*

Elastic modulus	572 MPa
Poisson's ratio	0.45
Yield strength	12.7 MPa

The model was modified in that rolling movement and tangential loading at the surface (as calculated in the previous model) were included to account for tractive forces generated on the tibial plateau. In order to save computer time, the femoral sphere (radius = 55 mm) was assumed to be rigid. The tibial component, directly adopted from Beard [34], was meshed with a total of 800, 8 nodded, plain strain isoparametric quadrilateral elements. It was constrained in all directions and the dimensions of the component were assumed to 10 mm in depth, 40 mm in length, and 20 mm in width.

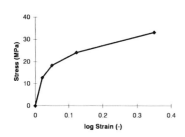

Figure 5.6: Stress-strain curve for UHMWPE in compression [36]. A multilinear approximation was used to model the material properties of the tibial plateau.

The normal load was applied to the bottom side of the tibial component as an uniformly distributed pressure across the tibial length. Since ADINA does not allow the addition of tangential forces in the contact zone, the tractive forces were incorporated using a friction coefficient and a minimal displacement of the femoral condyle which was directed in the same direction as the tractive force. The friction coefficients were varied according to the ratio F_t/F_n as determined by the previous knee model. Rolling motion of the condyle on the tibia was included in a quasistatic manner and obtained by rotation and translation of the sphere nodes, assuming the same contact path as described in section 5.2.2. A sinusoidal algorithm allowed the directional change of rolling movement in the posterior portion of the tibial plateau to be included. Load and motion were applied in incremental steps (20 N or 0.02 mm) to account for the non-linearities of the polyethylene component.

Normal and tractive forces at 64 distinct contact points were obtained as input from the previous model. The stress analysis of the tibial insert was performed for ten distinctive contact load situations on the polyethylene surface. Contours of the maximum principle stresses and maximum shear stresses were plotted over the entire cross-section of the polyethylene insert.

5.3 Results

5.3.1 General Force Pattern for Normal Gait

While the magnitude and pattern of the compression force $\bar{F}_n$ of Figure 5.7 was similar to previously published results [1,3], the variation of $\bar{F}_n$ and $\bar{F}_t$ along the tibial contact region (Figure 5.8) was of significant interest. In the anterior region of the tibial plateau, $\bar{F}_t$ reached a peak, producing a posterior pull on the tibial surface. A second peak occurred in the posterior region as the femoral condyles rolled backwards with a relative angular acceleration as high as 50 rad/s^2. In contrast to sliding, the direction of the tangential force during tractive rolling is not necessarily in the direction of relative motion between first and second body (Figures 5.7, 5.8), since *accelerated* and *decelerated* conditions are possible. A reversal of the tractive force occurred at the posterior end of the contact region. The knee rolled forward with knee extension following midstance flexion and $\bar{F}_t$ did not change its postero-anterior direction until the end of stance.

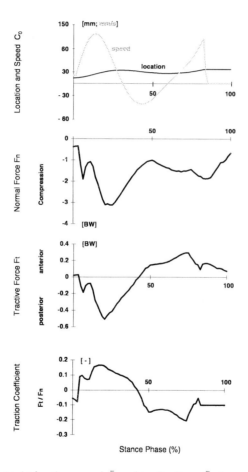

Figure 5.7: Contact point location, normal $\bar{F}_n$ and tractive Force $\bar{F}_t$ for a normal walking pattern (0% antagonistic muscle activity). Note the biphasic shape of the traction force with its sign change around midstance. The latter is independent of the direction of relative motion. At about 85% of stance, rolling motion comes to an end and sliding takes place. The traction coefficient F_t/F_n between femoral condyle and tibial component remains between ± 0.2 during normal walking.

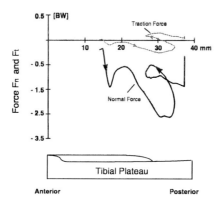

Figure 5.8: Sagittal view of the polyethylene component with loading history (normal gait pattern, no antagonistic muscle activity). The normal and traction force are shown with respect to their location on the articulating surface. Note that some tibial locations in the posterior region are overrun three times by the femoral condyles. Also the reversal in the direction of tractive force occurred in the posterior portion of the bearing.

5.3.2 Influence of the Coefficient of Friction

The coefficient of friction had a substantial effect on the generation of the tractive force $\vec{F}_t$. During rolling of the femoral condyles on the tibial plateau, the transfer of $\vec{F}_t$ can only occur below the limiting friction μ_s, otherwise sliding takes place. The traction coefficient F_t/F_n varied between ±0.2 (Figure 5.7). Thus, tractive rolling was possible during most of stance phase when $\mu_s = 0.2$, since tractive rolling occurs only when $|F_t/F_n| < \mu_s$. In late stance the knee flexion angle, α, exceeded 18° and sliding was initiated with a dynamic coefficient of friction of $\mu_d = 0.1$. As a result the tangential force was reduced.

Only a slight decrease from $\mu_s = 0.2$ initiated a transition from rolling to sliding during early midstance in the anterior region of the tibial plateau, while the occurrence of a local coefficient of $\mu_s < 0.12$ in the posterior region of the tibial plateau forced the knee to slide as the knee extended. The sudden transition from tractive rolling to sliding had implications for the dynamics of the knee. Sudden slip in the anterior region of the tibial plateau caused an anteriorly directed pull of 0.4 BW at the tibia, while slip in the posterior region, during knee extension, created a posteriorly directed force.

For all subsequent analyses a static coefficient of $\mu_s = 0.2$ was assumed, allowing stable tractive rolling during midstance and a smooth transition from rolling to sliding at the end of stance.

5.3.3 Influence of Tibial Conformity

While the distribution of the normal force along the tibial surface was similar for all conformity ratios, the magnitude of the tractive force changed with the curvature of the tibial plateau. The lowest traction force was found at a conformity ratio of 0.31 leading to a 30% reduction in the tractive force as compared to a flat surface (Figure 5.9). At conformity ratios beyond 0.35, femoral rollback was prevented because the traction coefficient exceeded limiting friction in the anterior portion of the tibial plateau and slip occurred.

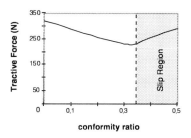

Figure 5.9: The tractive force is reduced with increasing conformity between femoral condyles and tibial component. Rolling motion will be inhibited if the conformity ratio is increased beyond 0.35.

5.3.4 Influence of Gait mechanics

There was a substantial change in the characteristics of the traction force when the gait characteristics were changed to the "Quadriceps Avoidance" gait pattern. The posteriorly directed traction force in the posterior region of the tibial surface was reduced (Figure 5.10). The reduction was primarily a result of the reduced quadriceps activity, diminishing the pull of the patellar ligament. The anteriorly directed traction force was not affected by the different gait characteristics.

An increase in antagonistic muscle activity in the quadriceps group had the largest influence on the peak traction force on the anterior portion of the tibial surface for the normal gait pattern (Figure 5.11). Even a 10% antagonistic level doubled the magnitude of the initial peak of the traction force. Only the initial tractive pull at heel strike was affected by antagonistic extensor muscle work. Antagonistic flexor muscle activity occurring later in stance had only a small effect on the tractive force. While the precise level of the antagonistic muscle activity is not known, in normal gait, levels up to 45% of antagonistic muscle activity have been considered; higher levels resulted in the occurrence of gross sliding ($\mu_t > \mu_s$). For the quadriceps avoidance gait pattern the influence was similar.

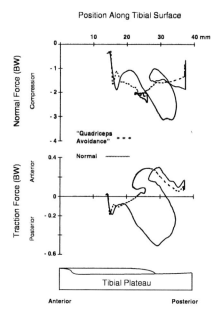

Figure 5.10: Sagittal view of the polyethylene component comparing the loading history of the normal (solid) with the "quadriceps avoidance" (dashed) gait pattern. Note that the antero-posterior pull in the posterior portion of the implant is missing for the quadriceps avoidance pattern. Also the normal force is reduced in that region.

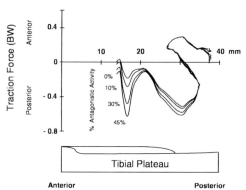

Figure 5.11: Antagonistic muscle activity produced the largest increase in tractive force in the anterior portion of the contact region.

5.3.5 Influence of Patella Position

A change in the initial patellar ligament angle (from 22 to 30 degrees) increased the second relative maximum of the traction curve. The results for the 'normal' gait pattern (0% antagonistic muscle activity) are illustrated in Figure 5.12. The second relative traction peak increased by about 40% with the initial patellar ligament angle increase for the case of normal gait. Changing the initial patellar ligament angle had no influence on the peak traction force for the 'quadriceps avoidance' gait, since the second relative traction maximum was absent for this gait pattern.

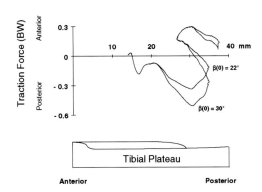

Figure 5.12: The initial patellar ligament angle at full extension influences the second peak of the traction force. The posteriorly directed pull is reduced with smaller patellar ligament angles.

The combination of a normal gait pattern with antagonistic muscle activity[*] and an increased patellar ligament angle produced the highest tractive forces on the flat tibial surface and was chosen to represent the subsequent conditions in stress analysis.

5.3.6 Stresses in the Tibial Component

The results of normal compressive stress and effective von Mises stress agreed well with those of Beard [34], which indicated that plastification of the polyethylene occurred in the posterior portion of the tibial plateau. Also the length of contact was similar with 3 to 6 mm. However, there was a notable difference in the location, magnitude and distribution of the maximum shear stress (σ_{yz}), since its asymmetric

---

[*] 30% antagonistic level was assumed for subsequent stress analysis (section 5.3.5)

character was directly affected by the direction of tractive force. As the tractive force was posteriorly directed from heel strike to approx. 45% of stance phase, the location of maximum shear stress within the contact zone remained posteriorly located (in front of the backwards moving condyle) until there was a reversal in the direction of tractive force, changing its position to anterior. During femoral rollback, the maximum shear stress moved from 1.3 mm to 0.6 mm towards the surface and became highest (7.4 MPa) in the posterior region (36.9 mm) of the tibial plateau at 21% of stance phase (Figure 5.13).

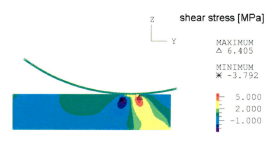

Figure 5.13: Shear stress contours in the posterior region of the implant (contact point located at 30.7 mm) at 21% of stance. Note the asymmetric character of the shear stress distribution.

While the maximum of the von Mises stress occurred below the surface (0.6 to 2 mm) at all times, it was interesting to analyze the normal tangential stress (σ_{yy}) at the surface since the tractive force will raise the tensile stress at the edge of contact. Immediately after heel strike, the tensile stress reached 7.3 MPa in the anterior portion of the implant (16.8 mm), while in the posterior region (26 to 38 mm) it fluctuated between 2.8 and 10 MPa. At the same time the compressive stress in tangential direction yielded -19.5 MPa (Figure 5.14).

At this point it seems appropriate to visualize the cyclic nature of the stress at certain locations of the implant. In a specific area of the posterior region (27 mm to 34 mm), the tangential contact stress changed periodically three times between tension and compression during stance phase of gait. In a similar

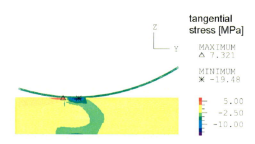

Figure 5.14: Tangential normal stress in the after heel strike (location 16.8 mm; 8% of stance). The tensional stress at the edge of contact reaches more than 7 MPa due to the traction peak immediately after heel strike (see Figure 5.11)

manner the shear stress underwent cyclic changes in direction, while the normal contact stress (σ_{ZZ}) exhibited a compressive to ompressive cyclic characteristic (Figure 5.15).

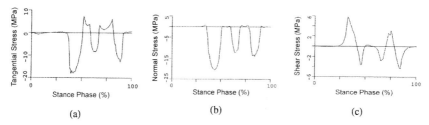

Figure 5.15: History of the tangential (a) and the normal (b) contact stress at one characteristic location (30.5 mm) during the stance phase of gait. Note the cyclic character of the stresses. Similar observations were made for the shear stress 0.75 mm below the surface (c).

5.4 Discussion

The purpose of this study was to test for conditions that can increase (or decrease) the tractive forces at the polyethylene component and, finally, consider the implications of these forces on the stresses within the tibial liner.

5.4.1 Coefficient of Friction and TKA Dynamics

This study suggests that there are conditions following TKA that can increase the magnitude of the tractive force acting at the knee. The coefficient of friction is probably the single most important factor influencing the magnitude of the tractive force at the tibio-femoral articulation. If the coefficient of friction increases, the articulation becomes sensitive to other factors such as gait or patello-femoral mechanics which then can also increase tractive forces.

The large variations in the coefficient of friction reported in the literature – a detailed review is given in section 8.1.2 – have been attributed to a variety of conditions. Davidson *et al.* [13] found that the dynamic coefficient of friction can start as low as 0.03, but rise to 0.1 after 200,000 load cycles. Similar observations have been made by others [12,38]. Many have attributed these variable coefficients of friction to changes in the interface, such as the creation of a polymer transfer film to the metal counterface. A change of the topology of the surfaces in contact can also have an effect on the interface friction as shown in section 8.3.3.

It should be noted that the coefficient of friction drives not only the magnitude of the tractive force but also the kinematics of a knee prosthesis. If a transition from rolling to sliding takes place, the rolling equilibrium will be disturbed and the knee joint becomes instable: a sudden arising unbalanced force component will push or pull at the tibia until it is compensated by soft tissue and/ or muscle forces. In the meantime gross sliding occurs. Depending on the overall contact conditions, the femoral condyles will slide independently from flexion/ extension anteriorly or posteriorly, possibly resulting in conditions which are clinically termed "negative femoral rollback" [39] This implies that the motion in a TKA will not be smooth and continuous as has been observed *in vivo* using fluoroscopic analysis[*] [40-42]. This is in contrast to that of a natural knee where the friction is negligible. Similar results have been described by Sathasivam and Walker [43] using a computer model with surface friction for the prediction of total knee kinematics. Their findings were useful in explaining the kinematic behaviour of different knee prostheses tested on a simulator [44].

While the aforementioned study [43] predicted widely different contact paths on the tibial surface depending on the condylar geometry, the current study suggests that not only the kinematics but also the magnitude of tractive force can be greatly affected by the design (i.e. conformity) of the prosthesis: a 30% reduction in tractive force can be achieved by applying the optimum conformity ratio.

5.4.2 Patient and Surgical Related Factors Affecting Surface Traction

An increased coefficient of friction will force the tibial articulation to sustain a larger traction force associated with different gait characteristics or variations in surgical technique. Large variation in the tractive forces at the knee can result from different patterns of walking. For example, the quadriceps avoidance type of gait reduced the tractive forces of the joint. Also a more flexed position of the knee at heel strike would be beneficial for the articulation, since it changes the slope of the patellar ligament to a more vertical direction, thus, reducing the tractive pull during quadriceps activity. Recently, it has been reported that gait affects the clinical outcome of tibial component fixation [45,46]. The patient group constituting the poor prognosis group had significantly larger peak flexion moments during level walking than the good prognosis group ($p < .005$). Also the mean value of the peak extension moments at heel strike was significantly larger ($p < .01$). Using the model of the current study, both

---

[*] It should be noted that there are still some limitations in the methodology of fluoroscopic analysis. Thus, the observed non-continuities during knee motion may reflect – in part – measurement errors also (personal communication with Prof. Andriacchi, Stanford University, USA).

conditions suggest high tangential forces at the tibial plateau with impact – either directly on the component fixation or indirectly due to wear particle generation – on the loosening process of the prosthesis.

Surgical factors may also influence tractive forces. The orientation of the patellar ligament will increase with thicker patella replacement. While there are other factors, like patella wear and breakage, that should be considered in deciding on the thickness of the patella, this study suggests that a thicker patella could increase tractive forces at the tibio-femoral articulation. It is important to note that this model focused on the generation of tractive forces under conditions of normal alignment. As has been shown in a study by Wasielewski *et al.* [26], rotational and varus malalignment are surgical factors which can lead to severe damage of the tibial component.

5.4.3 Limitations of the Knee Model

It is essential to reflect on the scope and the limitation of the methodology used to conduct this investigation. The analysis only considers the generation of tractive forces in the sagittal plane. No attempt was made to further investigate the distribution of those forces between the medial and the lateral tibial plateau. Femoral spin, the medio-lateral force component, and internal/ external moments can also contribute to the tangential forces generated at the knee joint. The choice of muscle lines of action, as well as the ligament approximation, influence the tangential component of the resultant force and, thus, may affect variations in the predicted tractive forces.

The assumed fixed kinematic linkage of pure rolling and sliding during knee flexion, based on passive knee kinematic studies, might vary from subject to subject. Furthermore, as shown in section 8.3.2, the maximum coefficient of friction does not occur during pure rolling but at a slide-roll ratio of about 40% to 50%.

The model assumed that the retained posterior cruciate ligament following TKA does not tighten earlier than 18° of flexion based on previous studies. A PCL that functions earlier in flexion could reduce the traction force along the articular surface which normally occurs later in the stance phase. The presence of antagonistic muscle activity alters both the normal and some components of the tractive forces. There was an effort to select the appropriate combination of agonistic and antagonistic muscles based on patterns of flexion-extension moments and EMG measurements from patients in the gait laboratory. The parametric approach of investigating the effect of antagonistic muscle activity [3] allowed the estimation of the contribution of antagonistic muscle activity.

Last but not least the input data from gait analysis may have errors by the time of implementation. As a primary artifact, marker movement due to skin motion has to be considered when evaluating internal movement of bony segments. The influence of these artifacts on the results and thus the limitation of has been described by Andriacchi and Strickland [59].

In spite of the model limitations, the model predictions seemed to be consistent with clinical outcome studies [45,46] as well as with retrieval analysis. As will be shown in section 6.3.2, the wear pattern and wear area suggest a rolling wear mode during articulation.

5.4.4 Tractive Rolling and its Implication on the Stress Conditions

As the femoral condyle rolls from anterior to posterior, elements in the path of the polyethylene become deformed and undeformed undergoing tension and compression (Figure 3.10). Thus, the moving contact area between the articulating surfaces causes portions of the surface to be subjected to cyclic stress which would provide the necessary conditions leading to fatigue of the polyethylene. While this effect has been stated already in the literature [19,47], it is interesting to look specifically at the posterior portion of the bearing, where a reversal of motion takes place and some elements of the polyethylene are passed over three times during stance phase. Assuming that a complete gait cycle lasts 1 second, a cyclic change of stresses at approximately 5 Hz (!) is generated. In other words, 1 million gait cycles will cause 3 million fatigue cycles, if tractive rolling occurs during midstance.

While surface fatigue cracks are usually associated with cyclic compressive-tensile stresses, Pruitt *et al.* [48] experimentally demonstrated that those cracks could be initiated and propagated also by fully compressive loading. The final length of the crack was dependent on the load ratio of the fatigue cycle: fatigue cracks propagated to greater lengths as the load ratio was increased. In this study it has been shown that both types of cyclic stresses – fully compressive and compressive-tensile – arise at the tibial plateau during normal walking.

With increasing tractive force, the location of the maximum shear stress was found to come closer towards the surface. This may be a necessary condition to initiate surface cracking [17], whereby it should be noted that the maximum shear stress did not rise above 0.6 mm below the surface. The coefficient of friction had to be increased to $\mu = 0.5$ to bring the maximum shear stress to the top of the material [49].

The limitations of the finite element model have to be considered. The quasi-static model was two-dimensional and, thus, stresses were analyzed in the sagittal plane only. It employed elastic material properties, although the UHMWPE articulation may be better described as viscoelastic component [50,51]. As shown in a recent study [52], the viscoelastic material properties lower the magnitude of the peak contact stresses and indicate a relation of contact area to rolling speed (hence, requiring a dynamic model). They do not, however, change the overall behaviour of the contact stresses as compared to elastic solutions.

Although wear and fatigue damage is associated with time, the loading history of only one gait cycle has been employed. As has been shown by Reeves *et al.* [61], the accumulation of stress residuals is dependent on the number of applied load cycles. The accumulated plastic deformation, caused by subsurface yielding of the polyethylene, leads to a region of compressive residual stress which is equilibrated by tensile residual stress at the surface. This tensile stress may combine with the tension generated by the femoral condyles at the edge of contact and contribute to accelerated fatigue damage [62].

In summary, the stress analysis confirmed that the fatigue life of tibial components may be shortened due to tractive rolling, especially when the same tibial contact spot is overrun several times during the stance phase of gait. However, the results did not illustrate the observation of Blunn *et al.* [53] that the highest *surface* damage of polyethylene occurred when normal forces acted coincidentally with tangential forces. Therefore, in the author's opinion, the analysis of the global stress condition does not appropriately portray the effective damage mechanism at the surface. The latter is supported by the clinical observation, that flat tibial plateaus do not wear faster than more conforming inlays [54,55], although there are numerous analytical and experimental studies which highlight the higher peak stresses in less conforming designs [47,56-58]. A precise, micro-scale analysis of the wear pattern of retrieved specimens might render additional information on the actual damage mechanism.

6 Early Wear Regime in Retrieved TKA Components

6.1 Introduction

6.1.1 Retrieval Analysis of Knee Replacements

As reviewed in chapter 4, wear of the tibial UHMWPE component – even in the absence of catastrophic failure – is a potential cause of mechanical failure in total knee arthroplasty (TKA). The analysis of retrieved inserts demonstrates the complexity of the problem [1-14]: many different changes in the appearance of the polyethylene bearing surface have been reported, including deformation, pitting, delamination, abrasion, burnishing, scratching, and embedded particulate. Usually these surface changes are referred to as "wear damage modes" and have been described in their severity using a subjective scoring system according to Hood *et al.* [1] *.

To improve the wear characteristics of tibial inserts, it is important to identify the prevalent wear mode and wear mechanism of the tribosystem, as outlined in chapter 3. Historically, much effort has gone into the interpretation of the wear features of the severe looking pits and delaminations, both considered as failure of the UHMWPE component [2,9,15,16]. The wear regime after UHMWPE failure, however, destroys the primary wear pattern, since the mechanical conditions (and hence the wear mode) under which the prosthesis was functioning have changed [17]. Particles generated due to pitting may lead to third-body problems, depending on the time they remain entrapped in the interface [18]. This effect may be more pronounced in more conforming designs [4]. Surface delaminations can change the general alignment of the artificial joint and disrupt any primary wear pattern. It has therefore been difficult to identify the early wear process during normal function from implants that have already failed clinically [19].

6.1.2 Early Surface Changes on Tibial Components

Recently, we reported the observation of subtle striations on the bearing portion of the tibial surface in a subset of components retrieved relatively early after implantation [20]. These inserts were all manufactured in the same company with a direct compression molding technique and had a flat surface (i.e. the articulation was non-conforming). It

---

* Using this methodology, the tibial surface is divided into different sections (e.g. anterior, posterior, medial, lateral) and each section is graded on a scale of 0 to 3 points for the presence and extent of a particular damage mode.

has been suggested that the specific appearance of the striated pattern is related to the contact mechanics of the knee and, thus, might play a key role in understanding surface disruption on the tibial articulation. In order to analyze whether this interpretation is correct or not it is necessary to gain further information regarding the occurrence, appearance and morphology of the *"striated pattern"*. Then, with additional knowledge from wear testing and an understanding of the kinematics and loading conditions ("operating variables") at the artificial joint, it may be possible to identify the actual wear mode and its underlying wear mechanism(s).

6.1.3 Purpose

The objective of this study was to investigate whether the *occurrence* of the striated pattern is independent from implant manufacturer and prosthesis design (i.e. flat/ dished, direct molded/ machined), whereas its *appearance* is design related. Further, on a retrieval subset with more carefully controlled variables, the relationship between the striations and the kinematics of a specific TKA design was tested. In order to gain a better understanding of the underlying wear mechanism, the macro- and micro-morphology of the striated area was analyzed, and its relationship with the contact mechanics of the particular design was examined.

6.2　Materials and Methods

6.2.1 Materials Selection

Forty-seven tibial components removed for reasons other than UHMWPE failure were studied. Specimens were excluded if third body particles (e.g. cement or metal) were present. The average time in situ ranged from 2 to 143 months (mean 33.2 months). Twenty-seven of the inserts were posterior cruciate retaining designs (Miller-Galante (MG I) or MG II) with a flat articulation and twelve components were posterior cruciate substituting designs with a dished polyethylene surface (Insall-Burstein (IB)). These 39 components were made by ZIMMER, the eight remaining retrievals were manufactured by BIOMET, DEPUY, HOWMEDICA, JOHNSON & JOHNSON, OSTEONICS, SMITH NEPHEW, and WRIGHT and of various design. The MG I -inserts (16 of 47) were directly compression molded with a smooth surface finish, whereas any other design had a machined surface finish[*]. Clinical data are listed in appendix 6.

---

[*] All seven MG II inserts were machined from bar extruded UHMWPE.

Table 6.1: Patient data for the MGI retrieval subset

Accession No.	Age at revision (years)	Sex	Reason for Removal	Retrieved Side	time in situ (months)
229	66	F	infection	*	2.5
282	64	F	malrotation	left	2.5
236	65	M	infection	right	5.5
002	68	M	instability	left	8
004	78	F	instability	left	13
237	62	F	pain	right	15
174	31	F	pain	right	19
003	61	M	patella subluxation	right	23
394	61	M	infection	right	23
001	55	F	pain	right	37
007	71	F	pain	left	40
S96-14630	58	F	femur fracture above TKA	left	143

*information not available

In order to gain a better understanding of the relationship between observed striated patterns and previously described contact mechanics of the knee (chapter 5) it seemed appropriate to form a subset of the retrieval collection described above. This subset consisted of twelve tibial liners of a single design (MG I) without visible damage by the unaided eye (e.g. starting delamination, severe pitting or scratching) and complete clinical record. For one of those implants (PT 237) the matching femoral condyles were available too. The MG I prosthesis has similar curvatures on the tibio-femoral articulating surfaces as mathematically modeled in the previous section. The tibial plateau is flat and unconstrained. The condyles of the femoral components are asymmetric with the distal radius of the lateral condyle larger than the medial condyle in the sagittal plane. In the frontal plane, the lateral and medial condyle represent cylinders and, thus, are unconstrained and highly conforming with the ratio of the frontal curvatures (femoral / tibial) equal to one. All UHMWPE plateaus articulated with titanium alloy condyles. Nine new, non-implanted MG I inserts of a single size and thickness (small; 7.5 mm) served as control. Important clinical data of the retrieval subset are listed in Table 6.1.

The MG I components had been manufactured by direct molding Himont-1900 resin [21]. To determine if the findings in micro-morphology were applicable to other base resins, two compression molded sheets of 1120 GUR and 1020 GUR (fabricated by Hoechst Germany) were also studied. While 1120 GUR contains calcium stearate, 1020 GUR is free of any additives.

6.2.2 Probe Preparation

Since the whole bearing surface was of interest for initial investigation, samples were not destroyed at the beginning. However, to place the tibial liners into the scanning electron microscope (SEM) probe chamber, they had to be cut through the elevated center, dividing the medial and the lateral articulating surfaces. Areas of interest were marked using a SEM sensitive pen and samples were mounted on an aluminum specimen holder with adhesive carbon tape. In order not to affect later analyses, no coating was applied to the samples.

Subsequently, the tibial liners were cut into rectangular blocks covering a surface area of about 8 x 8 mm. The cutting was performed using a water cooled band saw (Contact Point 300, EXAKT Apparatebau, Germany), while the tibial liner was fixed in an oscillating probe holder. The oscillation generated a moving contact point of the diamond coated saw blade during cutting. Thus, cutting artifacts due to plastic deformation and heat, which are common for line contact of the saw blade on UHMWPE, were absent. Out of eight blocks, microtome cuts of 300µm thickness parallel to the bearing surface were obtained. While most of these samples served to gain more detail of the previously mapped surface (thinner probes contribute to a higher resolution using Low Voltage SEM), one microtome cut was embedded in oil between two glass plates for confocal laser microscopy. The remaining blocks were used to prepare cross-sections of 50 or 200 µm for subsurface analyses. While the 50µm films were embedded in oil for transmission light microscopy or prepared to be melted in a heated microscopic stage (for inspection of impurities), the thick probes were directly fixed on a specimen holder for SEM analysis. For all those tasks a standard microtome (Jung Rotationsmikrotom 2055 Autocut, LEICA, Germany) was used at room temperature.

A peg covering 1 x 1mm of the bearing surface was cut out from a 8 x 8 mm block of one tibial liner (PT 004) for transmission electron microscope (TEM) analysis. The peg was sharpened to 0.1 x 0.1 mm under light microscopy control and exposed to chlorosulfonic acid for 5 hours to break up the amorphous structures of UHMWPE. Then, ultra microtome cuts of a thickness of 50 to 80 nm parallel to the bearing surface were prepared from the embrittled peg by room temperature (Reichert Ultracut S, LEICA, Germany). Using uranylacetate, samples were stained for 2.5 hours. This procedure replaces the previously gained sulfur with uranium within the amorphous regions of the polymer. Now the lamellae or other crystal structures can be distinguished from amorphous regions using TEM microscopy, since this method according to Kanig [23] contributes to an atomic z-contrast (the heavier uranium atom absorbs more energy than the lighter carbon atom).

6.2.3 Mapping of Surface Wear

All samples were first viewed by stereo light microscopy to exclude severely damaged[*] and third-body[†] particle containing components. Since polyethylene reflects light in a very diffuse manner, the subtle surface features of the samples were analyzed by light microscopy bright- and dark-field, as well as polarized modes (Orthoplan light reflection microscope, LEITZ, Germany). In addition, a video measuring system (SmartScope™, OPTICAL GAGING PRODUCTS INC., New York, USA)[‡], designed for high contrast surface imaging, served to digitize and record the outlines of the striated, polished (burnished), pitted and scratched regions of the worn surface. In particular, the striated morphological change at the surface of the polyethylene was analyzed. The observed striated patterns were divided into three categories: an *elongated striated pattern* (long striations with a length-to-width-ratio[§] greater than six, oriented in a single direction), a *short striated pattern* (short striations with a length-to-width-ratio less than six, direction of orientation apparent and angle between striations <30°) and a *randomly oriented pattern* (angle between two neighbouring striations ≥30°). Representative examples are plotted in Figure 6.1.

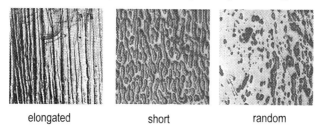

elongated   short   random

Figure 6.1: Representative examples of the three categories of striated patterns

Using AutoCad®, the area for each of those striated regions, as well as the area of any burnished, pitted and scratched region, was calculated from the digitized outlines (taken by the SmartScope™). The areas were calculated separately for the lateral and medial surfaces of every MG I component of the retrieval subset. Linear regression analyses were performed using SPSS 6.0 for WINDOWS. The probability level was set to 0.05. Time in situ as a dependent variable was correlated with areas of striations, burnishing, pitting, and the sum of these patterns for each component. In addition, the size of the medial area of each wear pattern was tested versus the size of the corresponding lateral area for each sample.

---

[*] i.e. incipient delamination, severe pitting, or scratching
[†] metal or bone cement
[‡] an in depth description of the system is available from von Lersner [55]
[§] quantitative description see section 6.2.4, section "orientation and spatial distribution"

6.2.4 Quantitative Analysis of the Surface Texture

Surface irregularities consist of a complex three-dimensional profile and loosing important information is inevitable when attempting to describe them by a single number, e.g. the average roughness R_a. For a more complete description, information is needed about the general surface contour and the topography of the surface asperities. The latter includes a description of asperity height, asperity shape and arrangement of peaks and valleys across the surface (i.e. spatial distribution and prevalent orientation of the asperities).

*Surface contour**

The video system was used to measure the surface contour of the retrieval subset and compare these to the unused control. The surface contours were measured to a z-resolution of 1.5 µm (lateral resolution 15 µm) over the bearing surface of the implant at 0.5 × 0.5 mm grid intervals. Each measured location was recorded relative to the bottom surface of the UHMWPE component and used to construct a profile map. The three-dimensional profile was used to analyze the relative geometrical differences within the component. A local coordinate system based on design landmarks of the respective components was used to accurately locate the observed surface features.

Asperity height and shape

Both, a laser and a stylus profilometer (Microfocus, UBM, Germany; and Perthometer S6P, MAHR, Germany) served to acquire qualitative and quantitative data of the z-topography of the striated pattern. While the laser profilometer (z-resolution 0.05µm; lateral resolution 1µm) was used to create three-dimensional topographical maps of six characteristic locations on the tibial plateau, the stylus profilometer (z-resolution 0.01 µm) was applied to interpret and summarize the topographical data of four individual specimens by means of surface roughness. Three different tracks per sample of about 5 mm were scanned in medio-lateral direction using a stylus tip of 5 µm radius. The average roughness of the profile (R_a), the maximum peak-to-valley height within the sampling length (R_{max}) and the so-called ten-point height (R_z), averaging five peak-to-valley values to reduce the effect of spurious irregularities, were determined (see section 3.3.1).

---

* A detailed protocol for the set-up of the measuring system can obtained from Seebeck [56].

Orientation and spatial distribution*

Areas of 2.5 x 2.5 mm on the tibial plateau – similar in location to those investigated by profilometry – were digitally recorded (mag. ~50x) using the SmartScope™ to determine the orientation and spatial distribution of the striations. The size of the areas was selected to gain a representative, homogeneous sample of striated texture. The recorded grayscale image was then transferred into a bitmap pattern using low-pass filtering and manual corrections of shiny spots, whereby the color "black" represented the dark appearing stripes and "white" any other part of the bearing portion (Figure 6.2).

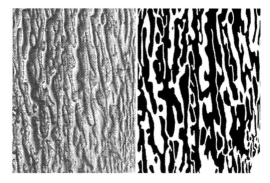

Figure 6.2: Transfer of the striated pattern (left) into a bitmap image

A customized software (*Star Length Distribution*) was used to quantify the orientation and spatial distribution of the striated pattern in all directions. This program was originally developed for the structural analysis of trabecular bone. The method and its validation have been described in detail by Smit *et al.* [24]. Briefly, a series of equally spaced grid lines was laid over the striated pattern under a certain angle φ (Figure 6.3a). Every pixel of a grid line located within a black portion of the bitmap pattern (and hence the striation) was taken as a sample point for the determination of the mean stretch of the striation in the direction φ.

This was done in 1° increments of φ from 0° to 180°. The global maximum of the plot "mean stretch vs. φ" was referred to as the *orientation* φ_{max} of the striated pattern and determined the (mean) *length* of the striations. The (mean) *width* of the striations was given at an angle perpendicular to φ_{max}. The same analysis was then performed for the white structures of the bitmap image. By adding the mean width of black and white structures, the *spatial distribution* of the striated pattern was determined. The amount of orientation of the striated pattern was defined by the *anisotropy* of the black structure which was determined by dividing the global maximum and the global minimum of the plot "mean stretch vs. φ" (Figure 6.3b).

---

* The spatial distribution describes the mean distance between two neighbouring asperity peaks.

In addition to selected size of area (here: 2.5 x 2.5 mm), resolution (pixel size) and spacing of the grid lines are important parameters to obtain a reliable description of the structure. Smit et al. [24] conducted an analysis of convergence for circular configurations and found that a sufficient accuracy (absolute error < 5%) is reached when the diameter of a circle covers 22 pixels. In this study the smallest structures occurring had a width of approximately 30 µm. Thus, a resolution of 1.4 µm was chosen. The selected area covered now 1730 × 1730 pixels, whereby a spacing of 10 pixels between every grid line yielded a satisfactory convergence and represented a reliable statistical sample of all points within the structure. In summary, the inaccuracy in spatial distribution was approximately 5%, while the error in orientation was even less than 0.5%, since the latter is determined relatively and not absolutely.

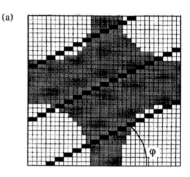

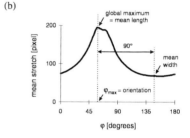

Figure 6.3: (a) Procedure to calculate the "star length distribution" (details see text). Reproduced from [57]. (b) Plot showing the mean stretch of the striations vs. φ. The global maximum determines the mean length of the striations, and perpendicular (90°) the mean width.

Surface hardness

The relative micro-hardness between peak and valley of the striated pattern was tested according to Vickers using a force of 0.25 N. This small force on the indentor allowed to evaluate hardness differences on a microscale measure.

6.2.5 Microscopic Description of the Surface

The microscopic characteristics of the striations were analyzed using a cold field emission scanning electron microscope (LVSEM: S-4500, HITACHI, Japan). Four samples, one machined, two molded and one control were viewed at magnifications up to 20,000x, using a minimum working distance of 7 mm at 1 kV accelerating voltage and 10 µA emission current. This relatively low accelerating voltage reduced the problem of charging artifacts, since the flow of entering (primary) and leaving (secondary) electrons is in equilibrium [25]. The LVSEM-image of the surface (generated by the secondary electrons) results in more topographical contrast as compared to standard SEM because of two reasons: firstly, the low accelerating voltage produces secondary electrons very close to the UHMWPE surface (0 to 50 nm) [26]; secondly surface details are not coated by a film of foreign material under which small differences of topography can vanish. Hence, using LVSEM techniques lateral resolutions on the order of 5 nm at 1.0 kV can be obtained for polymers [27].

The prevalent microscale texture (i.e. texture within the valleys and peaks of the striated surface pattern) was determined according to the protocol of section 6.2.4 using LVSEM images. Two characteristic 15 x 15 µm areas – one from the anterior and one from the posterior bearing portion – were analyzed to define orientation and spatial distribution of the occurring texture. The resolution was set to 0.01 µm / pixel for analysis. The z-topology of the microscale texture was quantitatively evaluated by means of white light scanning interferometry (Microxam, PHASE SHIFT TECHNOLOGY INC., Arizona, USA).

6.2.6 Determination of Impurities

In the past, considerable attention has been focused on inclusions in UHMWPE. They can have deleterious effects on the wear behaviour of the articulation by functioning as asperities (if present on the surface and connected to the matrix) or as third bodies (if liberated from the matrix). More recently, however, it has been demonstrated that most of the impurities which appeared as inclusions by light microscopy were in reality non-fused polyethylene powder particles [5,5,28,29].

Optical analysis by light microscopy

Therefore, in order to study the real physical state (inclusion or fusion defect) of apparent impurities, a technique according to Li [22] was applied. Microtome cuts obtained from the bulk UHMWPE with the presumed inclusions were put on a hot stage microscope. They were heated in 10° intervals up to 300°C and viewed under transmitted light, using the bright-field mode (DMRB transmission light microscope, LEITZ, Germany). The impurities should melt at a temperature of approximately 135°C (the melting temperature of UHMWPE [30,31]) and coalesce with the bulk material after cooling, if they are unconsolidated polyethylene. Otherwise the impurities are inclusions, composed of foreign material.

EDX-Analysis during microscopic investigation

The constituents of the inclusions were identified using energy dispersive X-ray spectroscopy (EDX: Link ISIS, OXFORD INSTRUMENTS, England). This tool is mostly applied in combination with SEM analysis, however, a minimum voltage of at least 10 kV is necessary to identify the constituent elements. As mentioned previously, this accelerating voltage would produce charging artifacts of the non-conductive polyethylene samples. Conductive coating, however, may reduce the probability of detecting light elements close to the surface. In addition, it may change the appearance of the SEM image which makes it difficult to re-find specific locations as seen under non-coated conditions. Therefore, inclusions on and under the surface of UHMWPE were viewed and analyzed using a variable pressure scanning electron microscope (VPSEM: S-3500N, HITACHI, Japan). This type of SEM dissipates surface charging due to the environment (air) at chamber pressures of 1 to 270 Pa. Thus, no sample preparation was necessary, although an accelerating voltage of 20 kV and chamber pressures from 30 to 70 Pa were applied for this study. The resolution of the image, however, was limited compared to LVSEM.

The applied voltage of 20 kV produces elastically back-scattered electrons (BSE) from the sample which allow the generation of BSE-images[*]. In difference to the SE[†]-mode, the BSE-mode contributes rather to material contrast than topographical contrast because the intensity of the BSE-detector signal is dependent on the atomic number of the element viewed. The higher the atomic number of the element in the periodic table, the brighter the BSE-image. Thus, local compositional differences are visible and now can be characterized using EDX.

Topography and hardness of the inclusions

Topography and relative hardness of the inclusions on the bearing surface of the bulk UHMWPE were determined using an atomic force microscope (AFM: D3000, DIGITAL INSTRUMENTS INC., USA). No specific sample preparation was necessary using AFM, however, it should be noted that full range in z-direction was 5μm limiting the analyses to smaller sized inclusions.

Van der Waals forces between the tip (ideally consisting of a single atom) of a silica cone and the surface of the sample are taken to produce the signal. The tip is fixed on a long cantilever arm to amplify the small inter-atomic forces (Figure 6.4). From several available modes (contact mode, lateral force mode, tapping mode, force

---

[*]Typically, a voltage of at least 5 kV is necessary in order to produce a sufficient number of electrons.

[†] SE = secondary electrons which are generated due to material penetration of the electrons of the primary beam

modulation mode), the tapping mode seems to be best suited for the analysis of polymer surfaces [32]. In this mode, the damping of the fast oscillating cantilever (20 nm at 500 kHz), which occurs when the tip hits the surface is employed. The relatively brief, intermittent contact between tip and sample prevents surface damage of the weak polymer. The actual topography of the sample is recorded analyzing the position of the cantilever which is controlled to produce a constant amplitude while oscillating ("height imaging"). "Phase imaging", which plots the phase lag between the oscillation drive and the cantilever

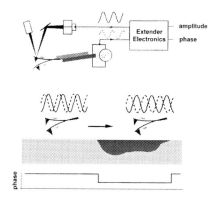

Figure 6.4: Principle of an Atomic Force Microscope. Phase imaging in the tapping mode employs the phase lag of the cantilever oscillation (solid wave) relative to the piezo drive (dashed wave). The cantilever answer is sensitive to the material properties at the surface

response, provides further information about variations in composition, hardness, friction, etc. of the material. Details about the analysis of polymer surfaces using AFM can be obtained from Reifer [32].

Relatively flat impurities (height < 4 µm) were investigated on the surface of the molded sheets. The striated wear pattern on the retrieved components restricted measurements to the "non-contact" area due to the height of the striations. An analysis of microtome cross sections would have been difficult also, since cutting marks can disturb the measurements and may lead to false interpretations [32]. It should be noted that the hardness is employed as a relative measure, since it is impossible to quantify hard contaminants in a soft matrix absolutely. The brighter the material's appearance in the phase image, the harder the properties of the material.

6.2.7 Analysis of Subsurface Characteristics

Determination of the polyethylene morphology

The morphology of the crystalline and amorphous polymer structures directly underneath the UHMWPE bearing surface were studied using a transmission electron microscope (TEM: CM200, PHILIPS, The Netherlands). The crystallographic structures found can be used as an indicator of the material's response to the stresses occurring during time in situ. As the samples were cut as thin films parallel to the surface, differences in depth were investigated viewing film by film.

TEM works in a manner similar to transmitted light, however, its resolution is largely enhanced. The by far shorter wave length of the applied electrons, in comparison to the wave length of light, accounts for the higher resolution. The electrons are emitted from a wolfram cathode and accelerated using a voltage of up to 200 kV before they hit the sample. Depending on the material's density and atomic weight of its composing elements the electrons are absorbed and scattered in a defined manner. These physical effects can be used to characterize the material's morphology using the bright-field (absorption) or the dark-field mode (scattering). While TEM can be easily applied for metals, polymers have to be dotted with foreign (heavy) elements to get sufficient contrast between the amorphous and crystalline structures (preparation technique see section 6.2.2). In addition, polymers are very sensitive to the electron beam and can be easily destroyed due to thermal problems.

Analysis of subsurface strain, cracks and cavities

Subsurface characteristics of polyethylene (e.g. cracks, cavities) are usually analyzed using cross-sections of the bulk material. However, it has to be considered that cutting itself may be a reason for these defects. The "white band", for example, which occurs in the cross sections of oxidized retrievals close to the surface [33], is produced while cutting the sample. The saw blade induces small cracks in the embrittled regions of the implant which causes the loss of transparency. By changing the sawing parameters from line to point contact, the embrittled regions are no longer present visually as "white bands" (personal communication: Biermann A., EXAKT Apparatebau, Germany). Also the analysis of residual strains within the cut sections is critical: plastic flow at the surface is produced during microtoming and cutting marks are easily visible. This might affect the results in using the birefringent technique on thin cut slices of UHMWPE. While Cooper *et al.* [34] tried to overcome these disturbing effects by a clear definition of the cutting direction, in this study a confocal laser scanning microscope (TCS 4D, LEICA, Germany) was utilized which allows subsurface analysis without cross-sectioning the material.

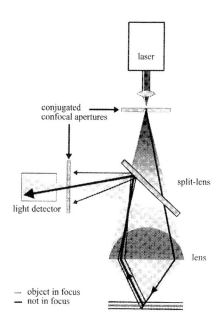

Figure 6.5: *Principle of a Confocal Laser Scanning Microscope. Objects outside the focus area are not recorded.*

The laser beam (*here:* wave length = 488 nm) transmitted through the bulk UHMWPE is scattered according to the optical properties of the material. The reflected signal can be now used to define the optical behaviour, whereby the confocal principle of the microscope confines analysis of the reflected signal to a certain level of material depth (Figure 6.5). In this protocol, the confocal conjugated apertures were set to study a layer of 1 µm thickness. By changing the focus of the reflected signal, layer by layer was investigated and information about optical differences in depth was acquired for a 125 x 125 µm area. Layers were plotted every 2.5 µm.

UHMWPE is optically isotropic, however, under deformed conditions it becomes anisotropic. The laser beam will be then reflected depending on the degree of anisotropy, and strains due to elastic or plastic deformation become visible. On free surfaces the beam is scattered distinctively, generating a more intense reflection which helps to differentiate (open) cracks or cavities from residual strains.

6.3 Results

6.3.1 Patterns of Wear on the Tibial Plateau

Typically, the retrieved components showed minimal visible surface changes. However, under high contrast or polarized imaging, striations on the surface were apparent emerging as contrast differences of bright and dark areas (Figure 6.6). More than 70% of the retrievals exhibited a morphological change in the surface of the polyethylene, which was not confined to a specific implant design or manufacturing company (Table 6.2). Polishing (or burnishing) occurred in nearly all components and was always adjacent to the striated pattern. Morphological characteristics which were interpreted as pitting were also found within all components (Figure 6.7). Mild scratching was also seen in most of the components and four MG I samples showed partial delamination due to patella failure. The components which did not show any striations were in situ a significantly shorter time than the retrievals with pattern (13.7±25.3 months vs. 41.3±36.3 months; p< 0.05), whereby there was no interaction of the manufacturer (ZIMMER/ non-ZIMMER), implant type, design (flat/ dished), nor surface finish.

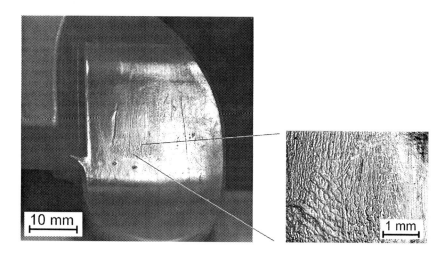

Figure 6.6: The typical appearance of the striated pattern on a tibial insert

Table 6.2: Components with a striated morphological change of the surface

manufacturer, type	design of articulation	surface finish	total number investigated	number with striated pattern	percentage
ZIMMER MG I	flat	direct molded	16	12	75%
ZIMMER MG II	flat	machined	11	7	64%
ZIMMER IB	dished	machined	12	8	66%
Others	-	-	8	6	75%

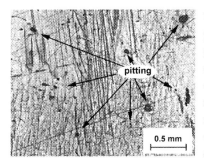

Figure 6.7: "Black dots" at the bearing surface, interpreted as pitting

Usually, all three categories of the striated pattern (i.e. elongated, short, random) were present on the component for all four design groups listed in Table 6.2. The occurring patterns in the flat MG I and II retrievals were located centrally on both medial and lateral plateaus. The striations were antero-posteriorly directed with the elongated pattern positioned mainly anteriorly to the short pattern. If present, these two surface characteristics were bordered by the randomly oriented pattern (Figure 6.8). In the dished IB components, the short and the randomly oriented pattern were predominant. Long striations, if observed, were shorter in length as compared to the long striations in MG components. In general, the striated pattern on the IB component was a lot more unstructured and less oriented than in the MG retrievals. Often, short and random pattern occurred concurrently within the same area (Figure 6.9).

In summary, the elongated pattern was more pronounced in flat retrievals and often missing in dished components (37% vs. 16%). In the latter group, the short pattern occurred mostly in conjunction with the random pattern. This difference in morphological appearance, however, was not yet significant (p>0.05) using chi-square analysis. It is likely that more components would raise this trend to a significant level.

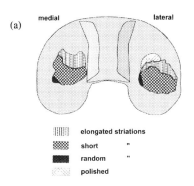
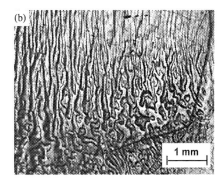

Figure 6.8: (a) Wear map of a typical MG-component. If present, the striations appeared centrally on both the medial and lateral surfaces. Usually the elongated pattern was in the anterior portion of the bearing surface and the short pattern in the posterior portion. (b) Transition of the elongated to the short striated pattern

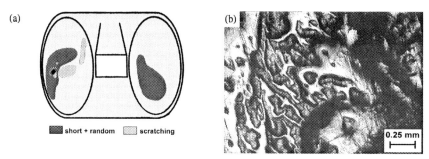

Figure 6.9: (a) Example of the observed wear patterns on the IB-component. Often short and random pattern occurred concurrently within the same area. (b) Typical appearance of the combined short and random pattern

6.3.2 Location and Area of the Striated Patterns

In the retrieval subset, i.e. the 12 "non-damaged" MG I components, 75% exhibited a striated morphological change of the surface. Here, the striated pattern was elongated and more oriented in the anterior bearing portion, while it was shorter and less oriented in the posterior part of the implant for nearly all samples. Using the orientation measures described in section 6.2.4 this qualitative description was confirmed quantitatively (Figures 6.10). It was interesting to note that not only the distinct antero-posterior orientation decreased in the posterior region of the implant (anisotropy posterior/anterior = 0.41), but also that the orientation of the striations was directed slightly to the lateral edge of the component.

None of the components had any striations near the medial or lateral edges, however, there were small wear regions at the anterior and posterior lips in 33% of the retrieval collection. If present, this surface damage was most often visible laterally on the anterior lip of the

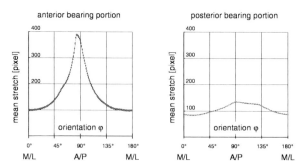

Figure 6.10: The striations in the anterior portion of the bearing had a distinct antero-posterior orientation (1 pixel ≈ 1.4 μm). The latter was reduced in the posterior portion of the implant.

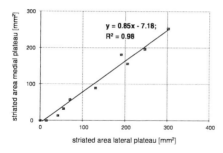

Figure 6.11: Regression analysis of medial versus lateral area of striations. The size of the wear areas on both plateaus was strongly related.

plateau. The average area of the striated pattern (186.0±198.2 mm^2) was larger than the area of the burnished region (45.4±63.4 mm^2). There was no correlation between any wear pattern area (or the sum of different pattern regions) and time in situ. However, the area of the striated pattern was smaller medially than laterally in the majority of the implants and there was a statistically significant correlation between both plateaus (adjusted r^2= 0.98, p<.0001). Regression analysis indicated a slope of 0.85 for the medial striated area as the dependent variable (Figure 6.11).

6.3.3 Surface Texture

Surface contour

It was not appropriate to calculate absolute deformations of the retrieval subset, since already new components of a single size varied more than 0.2 mm from labeled thickness. In addition, the surfaces of the new MG I components were not completely flat but deviated in central and lateral heights to form a smoothly dished surface exhibiting 0.15 mm depth on average. Despite those geometrical inaccuracies among the nine MG

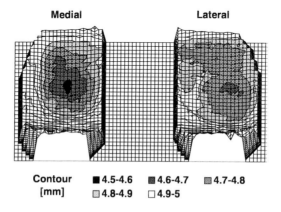

Figure 6.12: Contour plot (shown as thickness profile) of the same component as in Figure 6.8. Note that the biggest reduction in thickness is centrally located on both the medial and lateral plateau, while for this specific component a deeper impression was found on the medial side.

samples, plateau shape and thickness were comparable between medial and lateral side of the same, new component (thickness deviation at any laterally/ medially mirrored point: 0.023±0.047 mm). Thus, it was possible to quantify and locate *relative* differences for every retrieved component. Usually, the highest divergence was found within the central portion of the bearing area (Figure 6.12).

At this location, 7 retrieved components deviated more than 0.05 mm and up to 0.25 mm between the medial and lateral side, whereby 5 were deformed more medially than laterally.

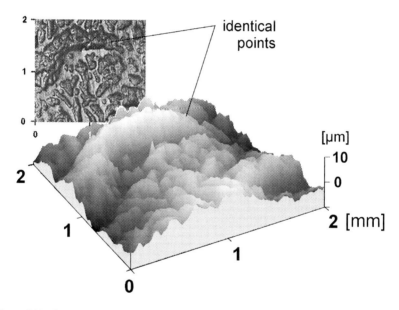

Figure 6.13: The striated pattern consisted of hills and valleys. Note that the topography of the structure is consistent with its visual appearance using high contrast imaging.

Orientation, height, shape and spacing of the striations

The topographical features of the worn surfaces were consistent with their visual appearances: the dark striations were located above the bright striations within the pattern (Figure 6.13). Thus, the whole pattern formed a landscape with antero-posteriorly oriented "peaks" and "valleys", whereby the peaks of the short pattern were usually higher than those of the elongated pattern. In general, the topology was less evident for the random pattern. The depth from peak to valley of a single asperity

Table 6.3: *Surface roughness measurements of the striated area of four retrieved samples using a stylus compared to an unused control.*

sample	R_{max} [μm]	R_z [μm]	R_a [μm]
retrieved	6.77±2.41	5.57±2.41	1.24±0.45
control	1.18	0.81	0.13

varied from 2 to 9μm, and averaging five peak-to-valley values (R_z) illustrated the morphological change at the surface when compared to the unused control (Table 6.3). It should be mentioned, that the profile shape of the striated pattern was remarkably different from the appearance of scratching which indicated much sharper peaks and valleys (Figure 6.14).

The spatial distribution (medio-lateral) of the striations was independent of the pattern category and determined to be 290±85 μm for various locations on the tibial plateau. However, there was a difference between the valley size of the elongated and the short pattern: the mean valley width of the elongated structure was smaller compared to that of the short structure (58 μm vs. 162 μm). It is interesting to note that these topographical features are mirrored in the orientation of the striations: the smaller the valley width the higher anisotropy of the striated pattern (3.9 vs. 1.6; Figure 6.10).

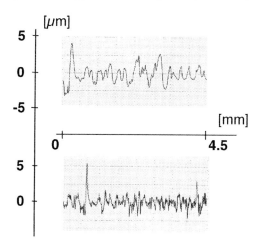

Figure 6.14: *The surface texture of striations (upper profile) was markedly different to scratching (lower profile). Shape and spatial distribution of the asperities differed between both wear patterns.*

Hardness

No relative differences between peaks and valleys of the striated pattern were found using Vickers micro-hardness analyses. Equally sized impressions of the cone were left on the polymer surface independent of testing location.

6.3.4 Microscopic Features of the Surface

At high magnification, using LVSEM, a row-like micro-morphology of the polyethylene was identified inside the valleys of the striated landscape which seemed to consist of micron-sized surface ripples (Figures 6.15). This presumption was verified using white light interferometry (Figure 6.16). The maximum height from peak to valley was 0.2 µm and the average roughness of these surface ripples was determined to $R_a = 0.041$µm. Similar to the macro-pattern there was a distinct difference between the anterior and posterior bearing portion: anteriorly (within the valleys of the elongated pattern) the row-like structures were highly oriented in antero-posterior direction (anisotropy = 7.5). The spatial distribution of those surface ripples could be determined to 0.95 µm (Figure 6.17a). Posteriorly, there was still this row-like morphology of the pattern, however, it exhibited fibrillar pull-out perpendicular to the micro-ripples. Thus, the pattern appeared less oriented (anisotropy = 2.2) with a spatial distribution of 1.5 µm (Figure 6.17b).

The hills of the striated landscape exhibited much rougher and less oriented surface features than the valleys. In addition to fibrillar pull-out, ridges perpendicular to the orientation of the striated pattern were found in the anterior bearing portion of the implant. These ridges were 5 to 20 µm in length and their elevation was in the sub-micron range as observed using white light interferometry (Figures 6.18). The ridges found in the posterior bearing portion (and, thus, occurring on the summits of the short striated pattern) were not as pronounced, nor as orientated.

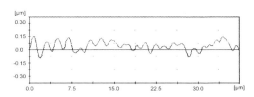

Figure 6.16: Roughness scan across the surface ripples in Figure 6.15a, iii

Retrieval Analysis 101

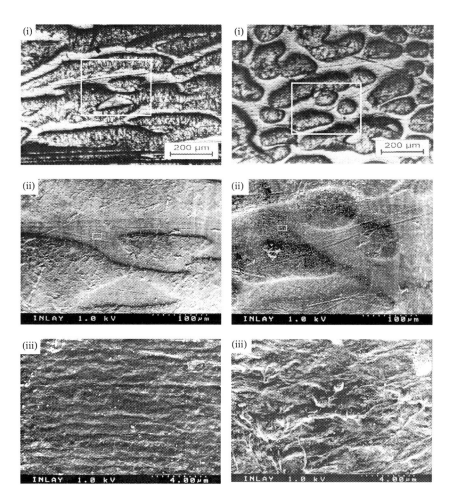

Figure 6.15a: The striated pattern in the anterior portion of the implant as seen by light microscopy (i), and LVSEM (ii, iii). Note the difference in the micro-morphology between bright and dark appearing structures (e.g. the white and black triangle in figure i). The 'bright' structures – forming the valleys of the striated landscape – consisted of micron-sized surface ripples (iii), which were oriented in the same direction (A/P) as the striations. The 'dark' structures exhibited a rougher surface morphology with fibrils and ridges perpendicular to the orientation of the pattern.

Figure 6.15b: The same picture-story for the posterior bearing portion. The surface morphology between anterior and posterior portion differed not only macroscopically (elongated/ short striations) but also microscopically: the surface ripples inside the valleys of the short were less oriented and exposed fibrillar pull-out (iii). These A/P-differences in the micro-morphology were only visible using high resolution imaging (compare to ii).

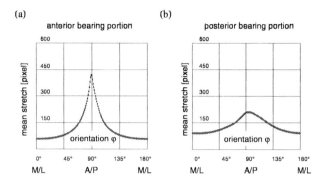

Figure 6.17: Orientation of the micro-ripples in the anterior and posterior portion of the implant (1 pixel ≈ 7.5 nm). Compare to Figures 6.15a/b, iii.

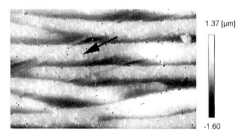

Figure 6.18: Height image taken by white light interferometry. Note the perpendicular ridges (arrow) on the summits of the striated pattern

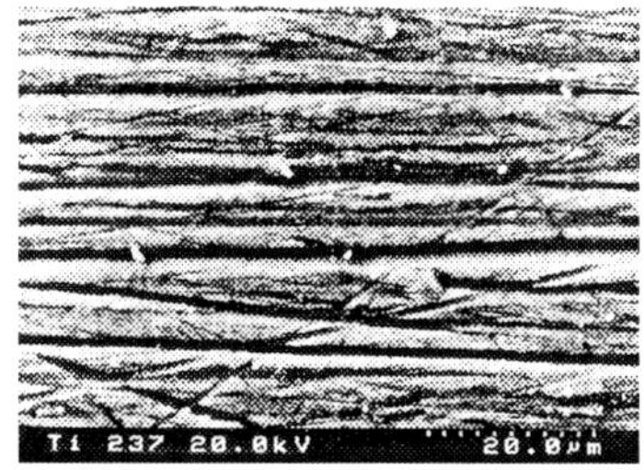

Figure 6.19: Scratches on the femoral condyle, mostly aligned in A/P-direction

The femoral condyles did not look damaged to the naked eye but appeared dull at locations where the implant had been in contact with the polyethylene counterface. This dead surface gloss was firstly assigned to polymer transfer onto the metal, however, closer examination, using the LVSEM, revealed fine scratches on the titanium surface which were responsible for the diffuse reflection of light. Similar to the wear patterns on the polyethylene counterface, those scratches were mainly antero-posteriorly oriented. The width of the deep scratches did not exceed 2 µm (Figure 6.19).

6.3.5 Impurities

Optical appearance and physical behaviour

The polyethylene samples indicated numerous, presumed contaminants on the surface as seen by light microscopy. These impurities looked similar on both retrieved liners and molded sheets (Figures 6.20 and 6.21).

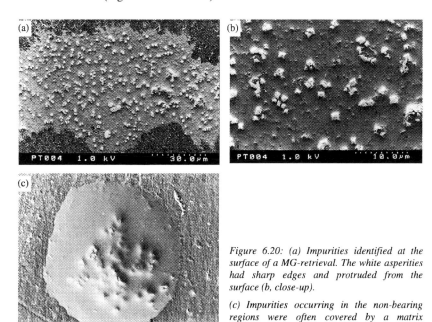

Figure 6.20: (a) Impurities identified at the surface of a MG-retrieval. The white asperities had sharp edges and protruded from the surface (b, close-up).

(c) Impurities occurring in the non-bearing regions were often covered by a matrix material different from bulk polyethylene.

In general, the impurity areas were of circular shape and of 20 to 50 µm in diameter, however, elongated areas with lengths of several 100 microns were found. Inside the impurity areas white asperities, protruding from a smooth looking matrix were identified using LVSEM. The asperities were 1 to 2 µm, rarely up to 10 µm in size. While most of them were of cubic shape, dendritic structures were observed as well. Still, it was uncertain whether these "bright" asperities were contaminants or sections of unconsolidated UHMWPE. The matrix surrounding the asperities (and defining the total width of the impurity) seemed to be different from polyethylene because of contrast and topography differences when viewing it by LVSEM. Sometimes the asperities appeared to be coated by this matrix material while others were not (Figures 6.20).

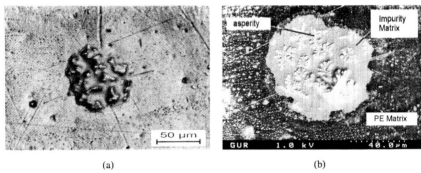

(a)   (b)

Figure 6.21: Light microscopy (a) and SE-image (b) taken from the same impurity at the surface of a molded sheet. Note the dendritic structure of the asperities inside the impurity matrix.

Figure 6.22: Cross-section through a retrieved tibial liner. The impurities were present throughout the depth of the bulk polyethylene

The impurities were found not only on the surface of the retrievals (or molded sheets) but also throughout the material (Figure 6.22). Interestingly, the enclosed asperities did not melt during in-situ annealing of the microtome sections. At 300°C they were still present, demonstrating that they were not unconsolidated polyethylene. The impurity matrix (i.e. the material between the asperities) was observed to undergo a transformation into the liquid state between 80 and 100°C, while the UHMWPE matrix melted at 140°C as expected (Figure 6.23). After cooling the asperities did not coalesce with the matrix.

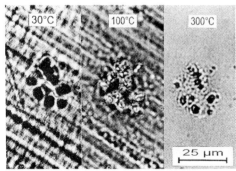

Figure 6.23: In situ annealing of polyethylene: the foreign material was still present at 300°C, whereas the UHMWPE matrix melted. The marks running from the upper left to the lower right are due to microtome cutting.

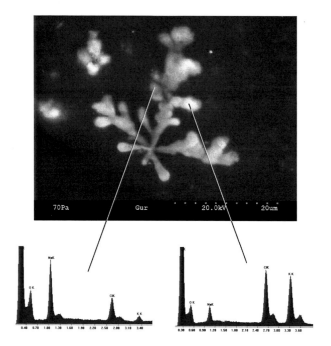

Figure 6.24: BSE-image of an impurity and EDX point analyses indicating the occurrence of sodium, potassium, and chlorine

Constituents

Unlike the SE-images (Figure 6.21), the impurity matrix was hard to identify on BSE-images, suggesting identical material constituents between UHMWPE and the matrix of the impurity. However, there was a high contrast between the asperities and the polyethylene. EDX-analysis confirmed these optical findings. The main constituents of the asperities were determined to be potassium and chlorine, while some of the asperities contained more sodium than potassium. If though, these asperity sections appeared grayish rather than bright white in the back-scattered mode (Figure 6.24). The EDX-spectrum of the material surrounding the asperities identified principally carbon and oxygen, while some sodium seemed to be amorphously distributed within the impurity matrix. The polyethylene at a certain distance of the impurity consisted of carbon and oxygen only. EDX surface mapping illustrates these impurities are salt minerals (Figure 6.25). It should be noted that the salt contaminants did not always appear crystalline, the amorphous state was also observed. Sodium, in particular, was often irregularly distributed within the whole matrix of the impurity.

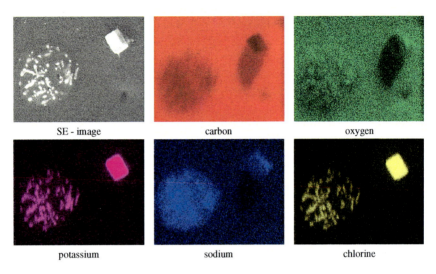

Figure 6.25: EDX surface mapping proved that the enclosed asperities are salt minerals which typically occurred in a cluster arrangement or occasionally as single blocks up to 10μm.

Topography and hardness

Figure 6.26 displays a three-dimensional view of a single impurity on a freshly molded GUR 1020 sample. Note the elevation of the impurity above the surrounding polyethylene matrix. The micro-topography of the impurity involves several, sharp-edged protrusions, coated by a very finely textured material. The protrusions represent the previously identified salt asperities, while the smooth coating portrays the covering features of the impurity matrix. Note the surface of the polyethylene matrix is much more "ruffled" in comparison to the impurity matrix.

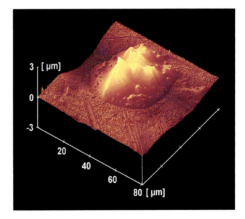

Figure 6.26: AFM image of a salt impurity found at the surface of a compression molded sheet. The protruding asperities were coated by a material different from polyethylene.

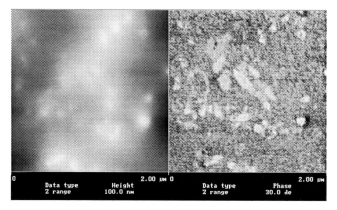

Figure 6.27: Height (left) and phase image (right) using AFM. Note that the asperities protruding from the surface are harder than the surrounding matrix (as verified using phase contrast)

Phase imaging using AFM demonstrated the differences in hardness, H, between the polyethylene matrix, the impurity matrix and asperities (Figure 6.27). The ranking has been determined to be as follows:

$$H_{impurity} < H_{UHMWPE} \ll H_{asperity}$$

6.3.6 Subsurface Characteristics

Morphology of the retrieved polyethylene

The ultra-microtome sections of UHMWPE, studied by means of transmission electron microscopy (TEM), showed crystalline and amorphous phases irregularly distributed throughout the material. Note that the preferred organization of the crystals, which can be identified as pairs of dark lines on the image, was lamellar (Figure 6.28a). The orientation of the lamellae varied depending on the examined location and sometimes was organized further to form spherulites. The average size of the lamellae, estimated from TEM micrographs, was about 20 nm.

In addition, layers of highly oriented nano-fibrils were found below the surface (Figure 6.28b). The development of these oriented layers could be roughly assigned to depth of no more than 5 µm, according to the number of ultra-microtome cuts performed (first 100 slices) and investigated. The spatial distribution of the nano-fibrils ranged from 15 to 25 nm which was again estimated from the micrograph. The fiber crystalline morphology suggested that plastic deformation occurred close to the surface.

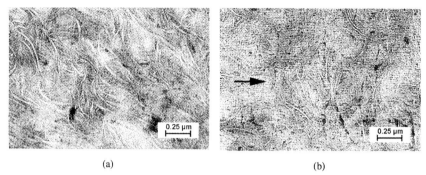

(a) (b)

Figure 6.28: (a) Transmission electron micrograph (orig. ×50,000) of a direct compression molded Miller-Galante component showing the lamellae as the preferred organization of the crystals. (b) Nano-fibrils were detected also, concurrently with the lamellae). The arrow indicates the direction of alignment of the nano-fibrils.

Residual subsurface strain

Samples of the bulk UHMWPE taken from one retrieved component did not show any cracks or cavities, only the occurrence of residual subsurface strain was observed.

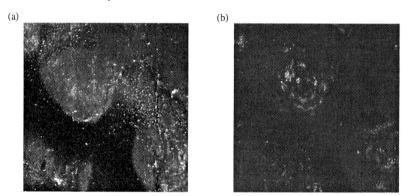

(a) (b)

Figure 6.29: Confocal microscope images (a) showing a 250×250 µm² region from the posterior tibial surface (elevated regions of the short striated pattern appear in gray) and (b) a plane parallel to that region in 18µm depth. Note that the residual strains occur underneath the summits.

Interestingly, the strain patterns underneath the surface appeared to replicate most of the geometry of the striated pattern at the surface. In general, strain concentrations occurred underneath the hills of the wear landscape up to a depth of 40 µm. (Figures 6.29 and 6.30).

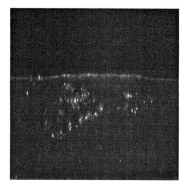

Figure 6.30: Cross-section (edge length of image 125 µm) through the striated area created by 25 planes of confocal images. Note the "flat" summit in the picture and the appearance of bright spots underneath. The latter indicate residual strains.

6.4 Discussion

6.4.1 Wear Pattern and Tibio-Femoral Contact Mechanics

There were several common features of the wear pattern seen in the early retrievals. The specific feature identified in the present study, is a striated morphological change of the tibial surface. Interestingly, there seems to be a relationship between the appearance of the striations (elongated, short or randomly oriented) and the amount of the constraint of the implant. The pattern is less oriented and considerably more split up in conforming implant designs. It appears that the early wear pattern provides a "fingerprint" of contact mechanics under conditions of normal function.

Indeed, a closer examination of the retrieval subset reveals that the consistent wear features of this type of implant can be related to the tibio-femoral contact mechanics of daily activities. During the stance phase of gait, the femoral condyles reciprocate between 0° and 20° of knee flexion under relatively high tractive and compressive loads at the tibial surface (section 5.3.1). The antero-posteriorly directed, elongated striations in the anterior portion of the implant suggest a tractive, unidirectional rollback without femoral spin. The shorter striations in the posterior portion of the implant refer to a mixed kinematic mode: rolling, sliding and spin seem to be superimposed in that area, while the rougher micro-appearance of this bearing portion may be related to the reversal of direction in tractive force occurring during late stance in that area (Figure 5.7). The bearing portions with the random oriented pattern (most often bordering the elongated and short patterns) may occasionally come into contact with the femoral condyles during activities other than level walking (e.g. stair climbing, raising from a chair, etc.).

Also the lateral to medial differences in wear characteristics can be interpreted with respect to the kinematics of the specific design characteristics of the MG knee system. The size of the striated areas on the medial and lateral plateau was strongly related with more wear laterally than medially. This is an interesting finding because one would assume more damage on the medial plateau due to higher normal contact loads occurring on this side. However, a closer look at the geometry of the MG femoral component provides an explanation: the radius of the distal portion of the femoral condyle is larger laterally than medially in order to reflect the anatomy of the natural condyles. This lateral to medial difference in the femoral curvature has been related to the normal tibio-femoral contact kinematics which takes place during flexion ($0°$- $20°$) [35]. During the stance phase of gait when rolling occurs, the larger lateral radius produces a greater movement on the lateral surface than on the medial surface. Thus, a larger wear area is generated on the lateral side.

Despite the fact that a larger wear area is produced laterally than medially, the components were more indented on the medial side in the majority of the implants[*]. This finding can be related to the smaller distal curvature of the medial condyle and/ or the higher contact loads occurring in the medial compartment of the implant. Both explanations are in agreement with the literature: Rose et al. [36] demonstrated that the depth of indentation increases exponentially with load, while Blunn et al. [37] observed deeper depressions in specimens loaded with femoral indentors having smaller contact radii.

6.4.2 Wear Pattern and Changes with Time in Situ

The occurrence of the striated wear pattern correlated significantly with time in situ. However, some long-term retrievals (e.g. one with 92 months) did not exhibit any striated pattern at all. Furthermore, there was no correlation with the area of any striated wear pattern (or the sum of it) and time in situ. In other words, the wear area did not increase with an increase in component life. This observation at first seems inconsistent with knee mechanics and associated wear. However, it has been shown in section 5.3.4 that the contact mechanics between the femoral condyle and tibial plateau are very sensitive to the dynamic characteristics of the knee of the specific patient. Gait adaptations, common to patients following TKA, may alter the kinematics and forces at the tibial plateau and, hence, influence the generated wear area. These results are consistent with published data;

---

[*]It should be noted that most of the indented volume is due to plastic deformation rather than material removal as has been recently shown by Walker et al. [39].

Wright et al. [38] did not find a correlation between implant duration and tibial polyethylene wear, such as pitting, burnishing and scratching, while delamination and deformation did depend on time. Wasielewski et al. [12] were not able to correlate wear severity and time in situ.

6.4.3 Relationship between Macro and Micro Surface Morphologies

Interestingly, at the microscopic level, the appearance of the worn polyethylene surface showed the same alignment features as the macro-pattern. The highly oriented ripples in the anterior portion of the bearing corresponded, both in direction and degree of orientation, to the elongated striations in that area. The rougher, less oriented micro-features in the posterior portion of the bearing coincided with the short striations. The occurrence of a ripple pattern on retrieved tibial components has been reported elsewhere [39,40] and, thus, does not appear to be exceptional, nor restricted to MG components. In a study by Walker et al. [39] the spacing of ripples on the surface of retrieved knee inserts was reported to be 1 to 5 µm. In this respect the finding corresponds with the results reported in this investigation, however, contrary to this study, the direction was assigned perpendicularly to the sliding direction and not in direction of motion. In a study by Wang et al. [40], ripples were found on the surface of UHMWPE hip cups and their spacing was smaller than 1µm. In certain areas of the articulating surface, fibrils grew perpendicularly from the ripples, similar the appearance of knee implants in the posterior portion of the bearing. It is suggested that the fibrillar pull-out is generated by the multi-directional motion trajectories occurring in the articulation of the hip cup [41] and in the posterior portion of the tibial insert.

The surface ripples, forming a corrugated micro-pattern, resolve the optical characteristics of the (macro) striated wear area on the polyethylene when viewed using light microscopy. The surface areas with oriented ripples reflect the light intensively, thus, resulting in a bright appearance, while the areas without micro-orientation cause a diffuse reflection of light and emerge as dark zones. It should be noted, that these ruffled surface areas occur on the hills of the striated area and are in direct contact with the femoral condyle. On the one hand this clarifies the occurrence of ridges and fibrillar pullout visible on the summits (Figure 6.18): fibrillar formation of polyethylene is a direct result of adhesion which occurs under two-body contact [42]. On the other hand it raises the question: what is the driving mechanism of the micro-ripple formation in the valleys of the striated landscape? Although the femoral condyles showed antero-posteriorly aligned scratches of the same order of magnitude as the micro-ripples on the polyethylene surface, the opposing metal surface appears not to be in direct contact with the corrugated micro-pattern.

6.4.4 The Occurrence of Salt Impurities in UHMWPE

A survey of related literature reveals that similar observations of wear morphology have been made under rolling abrasion as the active wear mechanism [43]. The latter would imply a third body involved, harder than polyethylene and harder than titanium. For example, bone cement particles, trapped within the articulation, would qualify for third-body damage [44]. The identified salt minerals in UHMWPE may also play a significant role as interfacial material. Once the particles are released from their matrix they aggravate third-body wear and can bridge the interface between the first and second body.

To the author's knowledge this is the first report about salt impurities in UHMWPE. Using the annealing experiment as suggested by Li [22], it was shown that the inclusions are not unconsolidated polyethylene particles. Furthermore, the salt impurities are not contaminants in the common sense. Those so-called "specks" are most often visible to the naked eye and consist of corrosion products (iron, chromium, nickel) from processing and catalyst/ additive residues (titanium, aluminum, calcium) from synthesis [31,45]. Sodium and potassium, however, are not acknowledged as foreign materials incorporated in UHMWPE. Table 6.4 lists the tolerated limits of foreign elements according to DIN 58834.

Table 6.4: Concentration limits of foreign elements in UHMWPE according to DIN 58834

foreign element	titanium	aluminum	calcium	chlorine
concentration limit	20 ppm	40 ppm	50 ppm	20 ppm

Since UHMWPE has been in clinical use for over 30 years and its wear properties, quality and morphological structure have been studied intensively, care must be taken in interpreting the "discovery" of these salt impurities. If the contaminants are in the nascent (virgin) powder as has been shown by the molded plates of GUR 1020 and GUR 1120, why are there no reports in the literature? The answer is most likely related to the utilization of high resolution imaging of UHMWPE without coating, an achievement made possible by recent SEM technology. In all likelihood the contaminants were overlooked when studied with routine high-voltage SEM as documented by Figure 6.31. Using gold as coating material, 20 kV acceleration voltage obfuscates the intricate surface morphology of the inclusion; the impurity matrix, surrounding the salt asperities, is no longer detectable and the salt particles themselves lose most of their characteristic topography. Thus, the salt asperities can hardly be distinguished from dust contamination which are common on SEM samples.

Because sodium, potassium and chlorine occur in human body fluids, artifacts resulting from the precipitation of chloride salts onto the surface of the retrieved device must be ruled out. However, this scenario is highly unlikely because these particles were also found on the surface of the directly molded sheets, as well as beneath the surface of the UHMWPE devices. Since salt is soluble in water, and all retrievals were washed using distilled water prior to investigation, the question arises why those particles were still present at the surface. Obviously their dissolution has not occurred, suggesting that they are coated by a material which does not dissolve in an aqueous environment. Indeed, the optical contrast between the impurity matrix and the polyethylene matrix, the results of EDX- analysis, the softer material properties and the lower melting point of the impurity matrix in comparison to the UHMWPE matrix all give reason to speculate that the immediate surrounding of the salt minerals consists of low-molecular weight hydrocarbons, e.g. different kinds of paraffin and/ or olefin. Thus, the salt asperities may be coated by a wax which would prevent their direct dissolution until the coating is rubbed away or the crystal breaks apart.

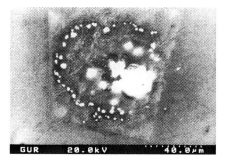

Figure 6.31: Most of the topographical contrast of the salt impurity in Figure 6.21 vanished using ordinary SEM technique (here: 20 kV acceleration voltage and gold coating).

6.4.5 The Role of the Salt Particles in the Wear Process

The striated wear pattern identified in this study was associated with areas of pitting and scratching. In the author's opinion, pitting and scratching are related to the occurrence of salt impurities (not only in this study!). As documented by our own previous work, initially there was a misinterpretation of the salt impurities using light microscopy [46]. Some of the impurities were falsely identified as pits due to a misleading light contrast. However, "real" pits at the surface of retrievals are also likely to be related to the occurrence of the impurities. Due to repeated loading under rolling contact – causing cyclic compressive and tensile stresses in the region of contact (section 5.3.5) – the origin of cracks at the solid surface is aggravated by material inhomogeneities [47]. Once initiated, the cracks propagate inclined to the surface along the border of the different material properties (e.g. impurity and polyethylene). The cyclic hydrostatic pressure in the synovia-filled crack will finally result in the pop-out of the softer paraffin/ olefin from the polyethylene matrix.

TEM examination of the morphological structure of the polyethylene just beneath the bearing surface revealed a microscopic cold flow within the surface layer of the polyethylene. Micro-creep occurring close to the surface of clinically retrieved inserts has been previously reported and related to "surface wrinkling and wear debris production" [33]. As documented by a study of Zhang et al. [48], the direction of the aligned nano-fibrils is consistent with the direction of tractive pull. Since these morphological features occurred very close to the surface it is suggested that they are generated due to occurring third bodies. It has been shown that entrapped cement particles can cause contact stresses leading to residual subsurface shear in UHMWPE [49]. The same behaviour may apply to any particle considerable harder than the polyethylene matrix and, thus, for sodium and potassium chloride minerals.

McNie et al. [50] conducted a finite element study to investigate the relationship between the geometry of an asperity (fixed on the counterface) and the plastic strain in the UHMWPE during a single pass of the asperity. In her model, the height of the parabolic asperity was set to 1 µm while its width was varied from 5 to 20 µm. It was found that the maximum von Mises stress increased linearly up to 52 MPa and the equivalent plastic strain increased non-linearly up to 49% plastic strain with decreasing asperity width. This compares with zero plastic strain produced on a smooth surface contact. These results demonstrate that salt particles in the range of 1 to 5 µm should be capable of causing plastic strain accumulation in the UHMWPE surface.

It should be considered that microtome preparation can cause flow of polyethylene and, thus, may introduce misinterpretations [34]. However, in this study the samples were etched prior to cutting which suppresses this problem. Furthermore, the nano-fibril morphology was only detected for distinct ultra-microtome sections, suggesting varying morphological characteristics of the polyethylene versus depth.

6.4.6 The Role of the Striated Pattern in the Wear Process

Using confocal laser microscopy, residual strains were not only detected within the surface layer of polyethylene but also further in depth (up to 40 µm) underneath the rounded summits of the striated pattern. Similar observations of strain concentrations in polyethylene have been reported by Fisher and Cooper [34,51]. It has been suggested that the development of localized residual subsurface shear strains beneath large asperities, ridges or peaks (amplitude approx. 10 µm) can cause subsurface failure and periodical increases in wear. Cooper et al. [34] speculated that for a smooth counterface this macroscopic wear of polyethylene asperities can even dominate the abrasive wear caused by the microscopic asperities of the metal counterface.

Not only local polyethylene asperities may rise contact stresses in the UHMWPE matrix, but also more expanded non-uniformity like machining marks [52,53]. In these studies of Bristol et al., contact stresses of machined versus molded UHMWPE knee inserts were evaluated by means of Fuji PresSensor® pressure sensitive films. The machining marks, which were topographically similar to the striations reported in this study (i.e. equal spacing, amplitude and shape), left grossly visible tracks in the Fuji film suggesting a large variability in the pressure across the contact region of the implant. It has been speculated by Bristol [53] that because of the reduced possibility of generating local transverse compression, abnormally high local shear stresses will be produced within the UHMWPE. Similar consequences may be associated to the striated surface pattern.

6.4.7 Wear History of the Tibial Plateau

In summary, two different levels of wear processes were identified: one characterized by a striated macroscopic change of the bearing surface and the other by the formation of microscopic ripples and fibrils. Both seem to be linked and become predominant after surface polishing (i.e. the machining marks have disappeared). These identified patterns of wear have a number of important implications in the analysis of damage of total knee replacement since they appear to be related to the specific dynamic characteristics of the implant. It is suggested that the salt impurities identified in this study aggravate the formation of the microscopic ripples and, thus, explain the close relationship between wear pattern and dynamics of the implant: once debonded from the matrix, the salt particles are ploughed through the interface on a path which is defined by the global kinematics of the first and second body. Therefore, the appearance of the ripple pattern differs between the anterior and posterior bearing portion. The transition from rolling to sliding and spin generates fibrils which grow most often perpendicularly from the ripples, most likely due to the combined action of cyclic micro-ploughing and interfacial adhesion [54]. In all likelihood, the accumulated plastic deformation of the surface due to micro-ploughing plus the overall stress conditions of the implant drive the generation of the macroscopic striations. The summits of the striated pattern become the predominant contact spots, reflected by the formation of ridges and fibrils due to adhesive mechanisms (film lubrication is reduced and boundary lubrication is more likely due to the occurrence of the striations). In conclusion, the characteristics of theses striations as well as their pattern of alignment provide important information in the interpretation of the wear history of retrieved implants.

7 Damage due to Tractive Rolling on the Tibial Plateau

7.1 Introduction

7.1.1 Wear Testing of Total Knee Arthroplasty (TKA)

Wear simulation is an important tool in the evaluation of the performance of the total knee implant under dynamic conditions. In the past, questions addressed have been mostly material or design related [e.g. 1-4], whereby kinematic issues have been investigated as well [5,6]. According to Walker *et al.* [6] the existing wear testing machines can be divided into two basic groups: the first group of simulators model the complete foot-shank-thigh linkage. In this case, the "muscles" are wrapped around the knee and balance the ground reaction forces which are applied underneath the foot. These simulators have been in use for over 20 years [7]. An advanced example of such a knee simulator has been developed by MTS Systems Inc. in collaboration with researchers from the Mayo Clinic and Johns Hopkins University [8]. The advantage of using this type of simulator is that the ground-to-foot reaction forces are well known and therefore highly accurate. However, the complexity of such a simulator limits the possibilities for long-term testing and, thus, for wear analysis.

The second group of simulators takes a more mechanistic approach towards testing, often by simplifying the true dynamics of the articulation of the joint. The reason for simplification is not only to reduce the costs and complexity of the testing device, but also to deal with the variability of joint forces, moments and motions. An example of a simplistic (but still appropriate) screening device is the "cylinder-on-flat" machine used by Davidson *et al.* [9]: a metal cylinder slides under reciprocating movement across a flat UHMWPE plateau to simulate the antero-posterior motion of an unconstrained knee prosthesis. Hence, pure sliding is applied while the line-contact between cylinder and UHMWPE sample generates high enough contact stresses to wear the material. More complex devices include the Stallforth [10] and Treharne [4] wear testing machines, which flex and extend the femoral condyle about a fixed axis. The "genuflector", developed by Paul *et al.* at the Massachusetts Institute of Technology (detailed description see [6]), uses a four-bar-linkage to generate antero-posterior sliding during flexion and extension of the knee prosthesis. While these early simulators are all movement-controlled, later models, including the Leeds knee simulator [11] and a recently introduced simulator by Walker *et al.* [6], take a combination of motions and forces as input to achieve the desired kinematics.

In spite of these efforts, the wear patterns observed in simulators do not completely reflect the kind of patterns seen in retrieved specimens. For example the severe delamination failures reported in some types of TKA [12] have not been reproduced in knee simulators [13]. Obviously, this problem is related to the definition and characterization of the relevant operating variables and parameters. None of the wear testing machines above took tractive rolling into account which might be substantial for a more precise replication of the specific contact mechanics of TKA.

7.1.2 Surface Damage and Contact Kinematics

Cyclic sliding has been assumed to be the most relevant kinematic action in the production of wear and – as a result – has been simulated in wear studies using a variety of test set-ups [e.g. 2, 9, 14, 15]. The dominance of cyclic sliding versus rolling in wear production was demonstrated by Blunn et al. [5] who modified a pin-on flat device to provide rolling and sliding under cyclic load. The femoral components were represented by cylinders with polished spherical ends of 25 mm and 75 mm radii which rolled or slid over a flat UHMWPE disc. Distilled water served as lubricant. To produce rolling, the "tibial" disc was reciprocated and the "femoral" cylinder was rotated in synchrony. To achieve sliding the cylinder was blocked. Rolling resulted in the generation of shallow wear tracks without major damage, while sliding produced deeper impressions with evidence of subsurface cracking. Blunn et al. [5] attributed the reduced amount of wear during rolling to the lack of frictional shear forces across the surface. However, as shown in the previous section (section 5.3.5), shear forces can not only occur under sliding conditions but also under *tractive* rolling conditions.

These tractive forces are generated by the combined action of muscle forces and acceleration or deceleration of the joint. The tractive forces act either against or in direction of motion as analyzed previously (section 5.1.2). Both conditions are likely to occur at the tibial plateau: as shown in Figures 5.6 and 5.7, the tractive force reverses direction during midstance of gait. The direction of tractive force may have important implications on the kinematics and the subsequent wear of the knee prosthesis.

7.1.3 Surface Damage and Third Bodies

Entrapped bone cement and metal particles have been recognized as potential third bodies. These materials are much harder than polyethylene and, therefore, may act as scratching indentors [16-19]. Hence, scratches and pits on the retrieved polyethylene components have often been related to cement particulates present. In section 6.4.5, however, it was suggested that even in the absence of cement debris third-body wear is likely to occur. Sodium and potassium minerals which are incorporated in polyethylene may aggravate *rolling abrasion* once these particles are detached from the matrix.

7.1.4 Purpose

The objective of this study was a) to investigate experimentally the influence of increasing tractive forces on the kinematics and early wear pattern of UHMWPE and b) to demonstrate that rolling abrasion (i.e. third-body wear) is likely to occur in absence of bone cement and metal debris.

7.2   Methods and Materials

7.2.1 Wear Testing Configuration

A special testing apparatus has been designed to simulate increasing tractive forces under conditions of pure rolling. The oscillating "Wheel-on-Flat" configuration consists of a driven cylindrical roller which is pressed against a flat polyethylene component under a constant normal load (Figure 7.1). The polyethylene component is mounted on a sledge and, hence, driven by the roller until an increasing tangential force, generated by a pneumatic cylinder, causes a transition from rolling to sliding. Wheel and sledge are then separated from each other and set back to the starting position to begin a new cycle. A photograph of the testing apparatus is provided in Figure 7.2.

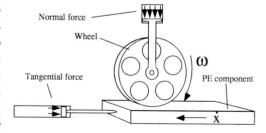

Figure 7.1: Mechanical concept of the "Wheel-on-Flat" apparatus: the sledge is driven by the wheel against a tangential force, actively produced by a pneumatic cylinder.

Figure 7.2: Photograph of the wear testing device

Test concept

As outlined in the introduction, the direction of the tractive force is not bound to the direction of motion of the contacting bodies. Although, *in vivo*, these conditions arise from complex interactions between muscular activity and external forces and moments, the tractive forces can be easily modeled by pushing or pulling the polyethylene plateau underneath the roller. The free body diagram of the polyethylene plateau, shown in Figure 7.3, aids in describing the associated conditions of equilibrium. Assuming, the sledge is driven at a constant velocity and supported by a frictionless bearing, the produced force F_{cyl} of the pneumatic cylinder is counterbalanced by the tractive force F_t

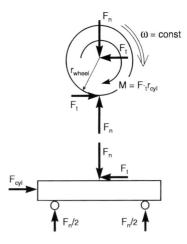

Figure 7.3: Free body diagram of the UHMWPE plateau and CoCr roller

$$\vec{F}_{cyl} + \vec{F}_t = 0; \tag{7.1}$$

These forces are in equilibrium as long as

$$\frac{F_t}{F_n} < \mu_{max};\tag{7.2}$$

where μ_{max} is the maximum coefficient of friction, applicable during rolling motion of the wheel, and F_n the normal force occurring at the contact. If μ_{max} is exceeded, sliding takes place between the wheel and plateau.

Obviously, magnitude and direction of the tractive force can be changed depending on the tangential force produced by the pneumatic cylinder. Tension or compression of the cylinder would generate tractive forces either in direction of the "condyle" movement ω or against it[*].

In order to keep a constant angular velocity ω and, thus, push or pull the sledge at a constant velocity v, the moment

$$\vec{M} = \vec{F}_t \times \vec{r}_{wheel};\tag{7.3}$$

needs to drive the wheel (nomenclature see Figure 7.3).

The control algorithm for the simulator is described conceptually by the flowchart shown in Figure 7.4. After the desired normal force is reached, the wheel drives the sledge against the tangential working cylinder at a constant velocity. The tangential force is then increased linearly using the displacement signal of the sledge as an input for closed loop control (Figure 7.5). Thus, a linearly increasing tractive force is generated at the articulation. Depending on the type of force within the cylinder (tensile or compressive), both directions of tractive force can be modeled. Thus, the cycle starts with free rolling (i.e. pure rolling without tractive force) of the wheel on the polyethylene component. It becomes more and more tractive due to the increasing tangential force. Subsequently, sliding occurs at the articulation. The cycle is then stopped, the wheel and the sledge are separated and set back to the starting positions to begin a new cycle. If for any reason sliding does not occur, the applied forces are removed when the sledge or the roller has reached its maximum excursion.

---

[*]According to the set-up of the local coordinate system in Figure 5.3 tractive forces are always described on the tibial plateau.

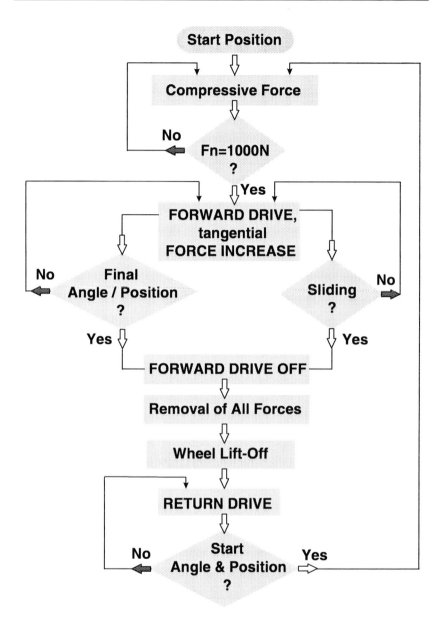

Figure 7.4: Concept of the control algorithm for one cycle during wear testing. The wheel starts its progression from the starting position under force control until gross sliding occurs or the endpoint at the plateau is reached. Then all actuators are shut off and the wheel is set back to its starting position under displacement control. The flow is identical for the tensile and compressive mode.

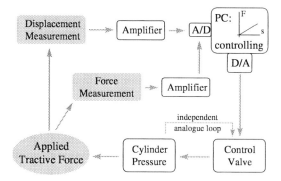

Figure 7.5: Schematic of the force control loop within the control algorithm. The actual displacement value of the sledge is used as an input for the desired force value (after mathematical modulation).

Sensors

The overall process depicted in Figure 7.4, i.e. the forces and displacements produced by the actuators, is controlled digitally using a custom made software (source code Turbo-Pascal®) on a PC. An independent analogue loop provides fine tuning for every actuator. An angular position sensor at the wheel, as well as a linear displacement sensor (both potentiometric) at the sledge and two force sensors record the magnitudes of load and displacement produced by the actuators (Figure 7.5). Important technical data and manufacturer addresses of the sensors are listed in appendix 7-I.

Mechanical Configuration

The testing apparatus is designed to provide easy access to the cobalt-chromium wheel and polyethylene plateau during or after measurements (Figure 7.6). A pneumatic short-stroke cylinder, driven by an 3/2 electric directional control valve and a pressure regulator, supplies the normal load in order to press the wheel onto the plateau. The wheel itself is driven by a pneumatic rotary actuator. Both (wheel and actuator) are guided in a frame which allows z-motion only. This arrangement eliminates malalignment of the cylindrical wheel on the polyethylene surface and, in addition, compensates irregularities within the articulation. A tensile spring supports the weight of the free-hanging unit (wheel and rotatory actuator). A high precision valve, designed for closed loop control, adjusts the air pressure on that rotary cylinder. Two 3/2 directional control valves on each side of the double-acting cylinder are additionally mounted for quick air release from the system. The same arrangement of valves is applied on the linear cylinder which provides the tangential force on the sledge. High precision bearings underneath the sledge account for negligible friction forces: 10N during motion initiation and 3...5 N under dynamic movement (based on 1000N normal load). Details on the pneumatic circuit are documented in appendix 7-II.

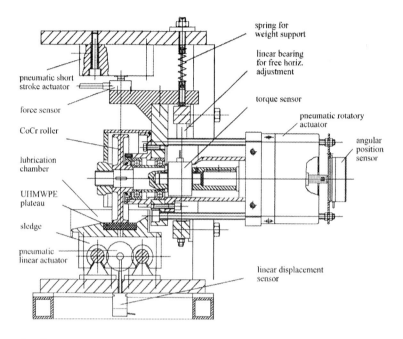

Figure 7.6: Cross-sectional view of the "Wheel-on-Flat" apparatus

7.2.2 Testing Protocol

The wear tests were conducted applying a constant normal force of 900N at the wheel. 900N reflect about half of the average normal load during stance phase of gait (based on the results presented in chapter 5), and were chosen because a single condyle was modeled in this wear test. After running-in, a linearly increasing tractive force across the "tibial" plateau was simulated. The length of the wear track covered at least 70 mm. This approach, namely a defined linear increase of tangential force plus the stretched wear track[*], made it possible to determine the precise loading history[†] for

---

[*] The wear track on retrievals with a flat articulation is typically 15 to 20 mm in length [24, 25]

[†] i.e. magnitude of tractive (and normal) force, number of applied load cycles, kinematic mode (rolling/ sliding)

every specific location on the polyethylene component and wheel. The magnitude of the applied force rates ranged from 1.1 to 2.3 N/mm (depending on the interface conditions, e.g. lubrication) and was kept constant for each sample after running-in. The wear tests were performed using both force modes: in the *tensile mode* the sledge was pulled underneath the wheel until sliding occurred; in the *compressive mode* the sledge was pushed. While the former simulates the (negative) tractive force occurring in the anterior bearing portion of the tibial implant, the latter models the (positive) force direction in the posterior region (see Figure 5.7). To generate damage at the early stage, a maximum of 0.5 million cycles were performed at 1 Hz. This represented an estimated implantation time of approximately 6 months. All tests were run at room temperature.

Throughout the experiment, the tractive force F_t (range ±500 N, total error < ± 0.1%), the normal force F_n (2 kN, ±0.5%), the displacement x of the plateau (100 mm, ±0.1%) and the motion α of the wheel (120°, ±0.05%) were measured. The data acquisition frequency was set to 150 Hz per channel and measurements were taken every 5000 cycles for an interval of 5 consecutive cycles. Thereby, the digital resolution was set to approx. 1% of the measured values. The maximum traction coefficient μ_t was used to describe quantitatively the transition from rolling to sliding. It was calculated from the determined maximum tractive force F_t during rolling movement according to:

$$\mu_t = \frac{F_t}{F_n}; \qquad (7.4)$$

where F_n is the normal force.

At the end of the test, metal wheel and polyethylene plateau were rinsed using distilled water to remove loose wear particles from the surfaces. Then, the specimens were dried for 24h and a measurement scale was applied to the polyethylene plateau for better orientation during microscopic analysis. The characteristics of the wear track on the wheel and the polyethylene component were analyzed using a light reflection microscope (Orthoplan, LEITZ, Germany) with polarized contrast. The results were compared to images of the same surface areas taken prior to investigation. In addition, a low voltage scanning electron microscope (LVSEM: S-4500, HITACHI, Japan) was used. No conductive coating was applied during the initial stage of investigation, in order to achieve comparable results to the retrieval analysis (sections 6.3.4 and 6.3.5). Later on, samples were sputtered with carbon to identify any occurring third-body particles using EDX-analysis at 15 kV (Link ISIS, OXFORD INSTRUMENTS, England). Any further steps of sample preparation were similar to the retrieval analysis protocol and details can be obtained from section 6.2.

7.2.3 Test Specimens and Conditions

5 compression molded UHMWPE components and 5 cylindrical wheels made of CoCr28Mo6, using one wheel on one component, were investigated. The polyethylene specimens were fabricated by ZIMMER INC., Warsaw, USA to allow direct comparison with retrieved Miller-Galante implants (Table 6.1). The metal specimens (outer diameter 100mm, width 20mm) were manufactured by IMPLANTCAST, Buxtehude, Germany. A polished surface finish similar to that of commercially available prostheses was applied (R_a= 0.04μm).

Three experiments were performed using the compressive force mode and no lubricant. This was done in an attempt to achieve worst case conditions and the highest tractive forces possible. Two experiments were carried out under lubricated conditions, whereby both modes – the compressive and tensile – were applied to study the achieved contact kinematics and the generated wear patterns. Distilled water, instead of bovine serum, was chosen as lubricant in order to provide a contaminant free environment without any potential third bodies included[*].

7.3 Results

7.3.1 Traction Coefficient

After each start of the cycle, rolling motion of the wheel was initiated without major tractive force on the polyethylene component. The tractive force increased as a linear function of sledge displacement, however, plotted as a time function, the behaviour was non-linear (Figure 7.7). This non-linearity was caused by (non-intended) deceleration of the wheel due to a increasing traction moment (equation 7.3) with sledge displacement. Once the difference between traveled distance on the wheel's circumference and produced displacement of the sledge increased beyond 6 mm, the cycle was stopped. The traction coefficient ranged from 0.13 to 0.17 and did not change significantly throughout dry testing. The transition from rolling to sliding was abrupt but did not cause any unstable reaction between wheel and sledge[†]. The sledge

---

[*]Although bovine serum is considered as the state of the art lubrication for wear testing of orthopaedic implants, distilled water has been shown to give similar wear results in knee simulators [13].

[†] In the contact analysis, evaluating the influence of friction (section 5.3.2), the transition from rolling to sliding was assumed to occur simultaneously with a sudden drop of the coefficient of friction (from static to dynamic). The application of this step function caused a mechanical instability which may affect TKA kinematics (section 5.4.1).

did not move further after sliding occurred, since the generated friction force and the compressive force of the cylinder were in equilibrium.

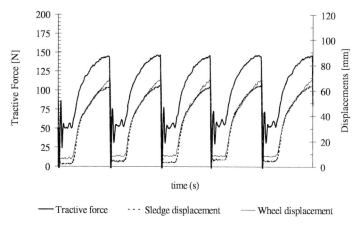

Figure 7.7: Five consecutive cycles showing the tractive force, sledge displacement and wheel displacement (calculated from angular progression × circumference) during compressive testing. Note that the velocity of the driving wheel was constant during the first 35 mm of displacement and then diminished. There was "pure" rolling for about 60 mm before sliding was initiated. A time delay relay, effective at each start of the cycle, provided low vibrations at the time of motion initiation. Otherwise the air filling of the cylinders would cause force oscillations.

7.3.2 Damage Caused by Tractive Rolling

Repeated rolling of the wheel across the polyethylene surface generated a (macroscopically visible) deformed path, mostly from shakedown of the component, while the metal wheel showed only minor surface damage to the naked eye: unidirectional dull stripes (up to 0.2 mm in width), running in the direction of motion on the bearing portion of the wheel were observed. For none of the wheels any influence of tractive force was visible, i.e. the stripes looked similar at all locations of the wear track on the metal surface. Using LVSEM, the stripes were found to be accumulations of subtle scratches, strictly aligned next to each other (Figure 7.8).

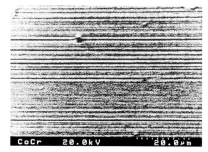

Figure 7.8: Scratches on the cobalt-chromium wheel strictly aligned in direction of motion ("A/P"). Note the equal spacing of the scratches

Their spatial distribution ranged from 0.9 μm to 1.5 μm (mean 1.24 μm), while their lateral width seldom exceeded 0.6 μm. The smooth appearance of the ripples suggested that they were generated due to micro-ploughing rather than micro-cutting. The frequency and proliferation of the scratches was surprisingly high, despite the fact that the experiments were performed under a clean environment and care was taken to prevent contamination with any foreign particulates.

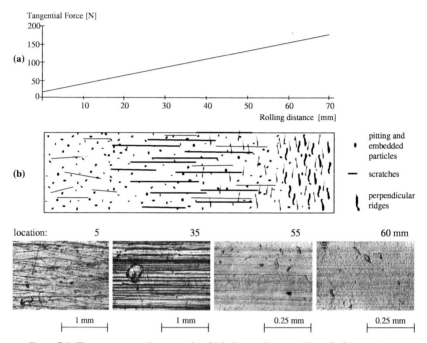

Figure 7.9: The appearance of wear on the tibial plateau changes with applied tractive force.

In contrast to the metal component, the severity of damage on the polyethylene surface increased with increasing tractive force (Figures 7.9). Over the whole wear track, pitting and re-embedded polyethylene particles were observed. The size of the pits was usually confined to a diameter of 10 to 20 μm, whereas some of the re-embedded particles gained several hundred microns in width. Longitudinal scratching also occurred, however, its appearance varied with location on the wear track. On the first 30 mm, random oriented scratches with 0° to 20° off-set from the direction of motion were visible. From 30 to 50 mm, these scratches were found to be more indented and precisely oriented in the direction of movement. In the last 20 mm of the wear track, the scratches disappeared and another mode of damage occurred. At that point, ridges perpendicular to the direction of motion came into view, pronounced in height and frequency on the last few millimeters of the wear track. In that area the ridges reached about 5 to 10 μm in length and seemed to be built up rather than separated from the surface (Figure 7.10). Figure 7.9 shows the wear appearances of the whole track generated due to tractive rolling. Note that the different damage modes were concentrated in certain areas of the plateau and, thus, were dependent on the tractive force acting between wheel and polyethylene plateau.

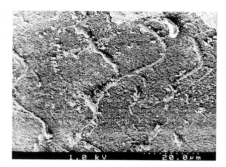

Figure 7.10: A close-up of the perpendicular ridges occurring at locations of highest tractive force at the tibial plateau

7.3.3 Influence of Tractive Force Direction on Kinematics and Wear

Figure 7.11: Parallel surface ripples (less organized than in Figure 7.12.) and polyethylene fibrils, generated due to sliding motion and water lubrication

Under distilled water lubrication, tractive rolling in the compressive mode (where the tractive force pointed against wheel progression) yielded unexpected results. During running-in, rolling took place with a traction coefficient of 0.04. However, after a few thousand cycles rolling motion came to an end and pure sliding was initiated right at the starting point. The wheel was caught in a small dip which was generated due to the impact of the normal force at the starting position. The friction was not sufficient to overcome these constraints. The subsequent generated wear pattern under conditions of pure sliding is shown in Figure 7.11: a row-like micro-morphology in the direction of movement was visible on the polyethylene surface. Parallel ripples exhibiting fibrillar pull-out were found next to some sub-micron particles.

These kinematic abnormalities were not encountered using the tensile mode under lubricated conditions. An evenly disposed shakedown of the polyethylene along the wear track did allow the wheel to proceed until the tractive force reached its maximum and forced the wheel to slide. The transition from rolling to sliding took place homogeneously but in contrast to the compressive mode, the relative displacement between roller and sledge was not terminated after sliding occurred. It had to be cut-off by the controller. The traction coefficient started at 0.04 to stay below 0.05 throughout the experiment.

7.3.4 Micro-damage due to Detached Salt Minerals

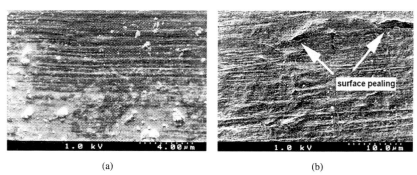

(a) (b)

Figure 7.12: (a) Longitudinal, antero-posterior oriented scratches on the polyethylene component caused under lubricated conditions in the tensile mode and (b) regions with surface pealing (arrows) at the far end of the wear track where sliding predominated over tractive rolling

The whole wear track of the polyethylene plateau was covered with fine, longitudinal scratches following the direction of motion. Using light microscopy, no differences in morphological appearance could be found along the course. However, morphological differences along the wear track were evident under the scanning electron microscope. Around the starting position, the wear area with the lowest tractive force applied, parallel ripples, strictly aligned in direction of motion were found. Their spatial distribution ranged from 0.3 to 0.6 µm (Figure 7.12a).

In the same area loose particles with varying sizes, hardly exceeding 1 µm, were found. In the middle portion of the wear track, where the tractive force reached 20-25 N, ripples were also evident. Their spatial distribution was about the same as in the starting section, whereby their indentation was more pronounced than in the starting section. Some of the surrounding particles appeared to have ploughed through the polymer, leaving micro-impressions on the surface. This morphological appearance of the surface did not change notably until the end of the wear track. In the section, however, where sliding became predominant the ripples vanished for the most part and another wear feature became apparent: the surface pealed off at different locations (Figure 7.12b).

After the samples were coated with carbon, the particles were analyzed. Their constituents were determined to be sodium, potassium and chlorine (Figure 7.13). In some cases, calcium and phosphor were identified, too (Figure 7.13). In general, the size and distribution of the particles was similar to that described for the retrieved Miller-Galante specimens (see section 6.3.5). Those which still stuck to the polyethylene matrix were 1 to 2 μm in size and generally occurred in clusters of widths ranging from 20 to 40 μm. Similar to the retrieved components, these clusters occurred not only at the surface but throughout the material. When detached from the matrix, the particles appeared to be slightly smaller and caught inbetween two adjacent ripples.

The wear appearances on the surface of the metal roller were not affected by lubrication. In all tests, the wheel showed similar wear patterns with no observed influence of the tractive force. No foreign material could be identified inside the scratches of the metal surface.

7.4 Discussion

To the author's knowledge, this is the first time that the appearances of surface wear were related to precise contact conditions occurring during rolling motion of the total knee joint. The approach, to control every load cycle by the relative location of contact between wheel and component (rather than time) allowed to classify the observed damage according to the magnitude of

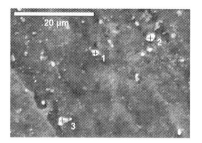

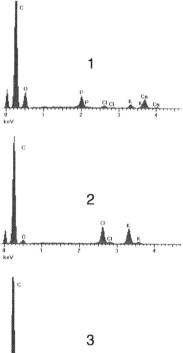

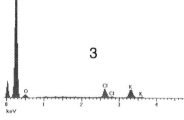

Figure 7.13: Third-bodies identified on the surface of the polyethylene component. Most of them were composed of potassium, sodium, and chloride. Particles containing phosphor (most likely from calcium stearate) were observed too, but to a less extent.

tractive force. In addition, it was possible to study the influence of tractive forces on the kinematic behaviour of wheel and component.

7.4.1 Effect of Tractive Forces on the Kinematics

As argued in section 5.4.1, the coefficient of friction between femoral condyle and tibial plateau is an important factor influencing the kinematics of the tibio-femoral articulation. Once the traction coefficient exceeds the maximum friction, gross sliding takes place. The direction of tractive force and the direction of condyle motion define whether the antero-posterior displacement comes to an end or is maintained under sliding motion. Under certain circumstances, e.g. when the femoral condyle is caught in local deformities on the tibial plateau, the occurrence of a tractive force in the direction of motion can initiate rolling, whereas a tractive force opposed to the direction of motion will hinder relative movement. Thus, the motion of a knee prosthesis will not be smooth and continuous, but rather will depend on the tractive force which is generated by complex interactions between muscular activity, external loads and the coefficient of friction (see section 5.3). These kinematic characteristics are in contrast to the natural knee where friction is 10 to 100-fold lower (section 5.1.1) and the tractive force negligible.

7.4.2 Damage Increase due to Tractive Force

Overall, the damage of polyethylene was determined by the magnitude of the tractive force. While there were different damage modes present on the polyethylene samples, surface pealing and perpendicular ridges were only present at the end of the wear track where the highest tractive forces occurred. The kinetics of the rolling wheel on the polyethylene component induced cyclic compressive-tensile stresses located closer to the surface due to the higher tangential loads (section 5.3.5). If these stresses exceed the fatigue strength of the polymer, they may cause the release of surface-near polyethylene layers. Some of these released layers are then sheared into the direction of tractive force and appear as ridges perpendicular to the direction of movement. It should be noted that partially released surface layers and perpendicular ridges are capable of forming particulate debris and/ or initiating progressive delamination. Both mechanisms are a potential problem in TKA.

Although scratching was identified along the whole wear track[*], its appearance changed depending on location (and, thus, tractive force magnitude), too. The

---

[*]In addition, pitting and re-embedded particles where observed along the whole wear track which occurred mostly at the non-lubricated samples

longitudinal scratches on the plateau were deeper and more oriented with increasing surface traction. They corresponded in appearance and dimension to scratches found on the wheel which indicates that they were generated by third-bodies. Great care was taken to prevent contamination from the articulating parts, hence, it is suggested that these third-bodies originated from inclusions within the polyethylene.

7.4.3 Rolling Abrasion due to Detached Salt Minerals

The inclusions on and within the testing samples were identified to be sodium and potassium chlorides, similar to those of the retrieved components (section 6.3.5). Since salt crystals are much harder than the metal and the plastic surface, it is highly possible that they act as indentors on the CoCr alloy and the polyethylene surfaces. It is suggested the detached salt particles induce a type wear which is known as *rolling abrasion* ("*Kornwälzverschleiß*") with abrasion being the predominant acting mechanism. The more ductile behaviour of the metal caused enhanced micro-ploughing on the metal surface compared to that on the more elastically reacting polymer (see equation 3.4 and Figure 3.9) . This difference in wear behaviour may disappear when sterilized and aged polyethylene samples are used. Due to the development of embrittled sub-surface bands [20-22] the latter will react more sensitive to plastic deformation.

Additionally to the use of non-aged polyethylene samples, another factor may explain why the ripple pattern was less apparent in the testing group as compared to the retrieval collection: micro-ploughing at the polyethylene surface is velocity dependent due to the viscoelastic/ viscoplastic behaviour of the material. Higher rolling speeds produce a stiffer but more elastic UHMWPE ("strain-rate hardening effect") [23]. Since the contact path was stretched to approximately 80 mm but still performed at a 1 Hz frequency, it can be speculated that the loading speed is inadequate to produce micro-creep.

Using distilled water as a lubricant, only the deep and heavy scratches vanished: a homogeneously grooved pattern in the sliding direction remained. It is therefore suggested that the salt particles are coated by a water resistant material (e.g. wax) because otherwise they should dissolve immediately in an aqueous environment. As soon as the coating is rubbed away or the salt crystal breaks into pieces, it will dissolve. The latter may explain why heavy scratches are missing when lubrication is used: bigger particles are more likely to get fractured and dissolved than smaller ones.

It is important to note that surface grooving occurred in the absence of bone cement or metal debris, thus, strengthening the hypothesis that embedded salt minerals are responsible for the production of the micrometer-sized ripples and scratches. These damage features have been observed previously using joint simulators. For example

Rostoker et al. [17] stated already in 1978: *"Despite the fact that the femoral head articulating with the polyethylene surface was highly polished and that one expects body fluids to contain suspensions of only soft tissue, there was a surprisingly high frequency and proliferation of scratches. [...] which would invoke the necessity of identifying the intrusion of hard particles since the surfaces themselves are not though capable of producing these effects."* While the aforementioned paper dealt with the wear of hip cups, Walker et al. [15] studied the deformation and wear of plastic knee components. His test set-up was quite similar to the author's study: a cobalt-chromium roller performed a ±30° oscillating motion on a partially conforming polyethylene component. Distilled water was supplied at a low drip rate and was not re-circulated. After the end of test the samples were investigated using SEM and the wear area was found *"to be grooved in the sliding direction"*. Interestingly, as seen from the SEM micrographs in [15], the ripples are in the same order of magnitude as compared to this study. Neither paper, however, offered an explanation for these wear patterns.

7.4.4 Limitations of the Study

At this point it may be appropriate to review the limitations and the scope of the study. Several factors influencing the tractive force and the damage of the polyethylene component have been not investigated. For instance, the effects of tractive forces were evaluated using a flat polyethylene plateau and a constant normal force. It has been previously shown that both variables influence surface traction (chapter 5) and contact stresses [26] of the polyethylene component and, thus, most probably wear. Water lubrication was used in an attempt to provide a clean environment, free of potential third-bodies. Although it was shown by others [13] that the (early) wear of the tibial polyethylene component is not greatly affected using water instead of serum lubrication, it must be considered that the use of bovine serum will change the corrosion mechanism of the metallic counterface compared to distilled water.

No quantitative wear measurements were performed, but the appearances of wear were classified according to applied tractive force and kinematic mode. This approach was taken in order to be able to relate wear patterns occurring on retrievals to the specific contact dynamics of the implant. At first glance, the applied number of load cycles may be considered as relatively low compared to other wear experiments in the literature. However, the aim of the study was to investigate surface wear at the *early* stage of the component's life.

7.4.5 Transferability of Wear Patterns to Findings on Retrievals

The early wear pattern due to tractive rolling has a number of important implications for the analysis of wear in total knee arthroplasty. It has been shown that *perpendicular ridges*, common to retrievals (see Figure 6.18), occur only under sufficient tractive load, while the appearance of *parallel* surface *ripples* and scratches follows the specific kinematic characteristics of the implant. Thus, it is possible to differentiate morphologically between rolling and sliding and relate the generated wear patterns to surface areas on retrieved tibial plateaus, where the same micro-features occur (see Figure 6.15 *iii*). While this is an important step in disentangling the kinematic and kinetic history of retrieved TKA implants, it was demonstrated that rolling abrasion starts to act relatively early on the component's surface. Hence, all subsequent occurring damage modes may be directly or indirectly affected by this type of wear. This is an important observation with respect to wear improvement of polyethylene as a bearing surface in total knee arthroplasty.

8 Analysis of Friction with Different Slide-Roll Ratios

8.1 Introduction

8.1.1 Rolling and Sliding of Knee Prostheses

In chapter 5, frictional conditions leading to pure rolling and pure sliding of the femur on the tibia were investigated and subsequently evaluated experimentally (chapter 7). Pure rolling and pure sliding, however, do not address the complete spectrum of kinematics that may occur in total knee arthroplasties (TKA). Typically, the articulation progresses as a combination of rolling *and* sliding (section 2.2.1) and, thus, different slide-roll ratios are achieved. The knowledge of the frictional characteristics for different degrees of combined rolling and sliding is important to appropriately address wear and kinematics of total knee implants.

The in-vivo variability of antero-posterior translation after knee arthroplasty has been studied by means of fluoroscopic analysis. Data of Stiehl *et al.* [1] have shown that squatting produces variable ratios of rolling and sliding between femoral condyles and tibial plateau during knee flexion. Similar results were obtained by Banks *et al.* [2], who investigated TKA motion patterns of patients walking on a treadmill. The analysis of their data demonstrated that the tibial plateau experiences non-linear condylar translations during knee

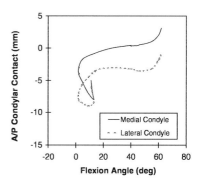

Figure 8.1: Contact path of a PCL- retaining knee arthroplasty. The path was determined from fluoroscopic analysis of patients during treadmill gait. Reproduced from [2]

flexion, indicating changing slide-roll ratios (Figure 8.1). A further experimental study [3] and subsequent theoretical analysis [4] from Walker *et al.* confirmed these characteristics: considerable variations of slide-roll ratios occur during knee flexion which are governed by the geometry and the frictional characteristics of the artificial joint.

8.1.2 Friction at the Articulation of Artificial Joints

Friction and lubrication regimes of artificial joints have been studied intensively over the last three decades. Most studies have focused on the total hip and, thus, pure sliding (and no rolling) was investigated. These experiments were usually performed using revolving pin-on-disc or reciprocating pin-on-plate testing configurations. To the author's knowledge, there have been only very few studies which have evaluated frictional properties of total knee replacement using category III or IV models (section 3.1.2). Moreover, pure sliding was used as kinematic input in these models [5,6]. Rolling motion should be considered because it changes the occurring lubrication regime: the development of fluid film lubrication is dependent on the relative velocity between solid body and counterbody [7]. Possible kinematic modes between the first and second body of the tribological system are summarized in Table 8.1.

Table 8.1 Kinematic modes of components in tribological systems

kinematic mode	velocities u_1; u_2	relative velocity $\rightarrow \Delta T_R$ $v_r = \|u_1 - u_2\|$	total velocity $\rightarrow$ EHD-film $v_s = \|u_1 + u_2\|$	slide-roll ratio $S = \left\|\frac{u_1 - u_2}{u_1}\right\| \cdot 100\%$
pure sliding	$u_1 > 0$; $u_2 = 0$	$v_r = u_1$	$v_s = u_1$	$S = 100\%$
rolling with slip ("wälzen")	$u_1 > 0$; $u_2 > 0$ $u_1 > u_2$	$v_r < u_1$	$v_s > u_1$	$0 < S < 100\%$
pure rolling	$u_1 > 0$; $u_2 = u_1$	$v_r = 0$	$v_s = 2u_1$	$S = 0$
rolling with slip ("wälzen")	$u_1 > 0$; $u_2 > 0$ $u_1 < u_2$	$v_r < u_2$	$v_s > u_2$	$0 < S < 100\%$
pure sliding	$u_1 = 0$; $u_2 > 0$	$v_r = u_2$	$v_s = u_2$	$S = 100\%$

The reported range of friction values of UHMWPE against metal is quite large. Soudry and Walker [6] analyzed the friction of total knees using cadaver specimens with implanted prostheses and water as lubricant. The detected friction coefficient varied between 0.021 and 0.041. Davidson et al. [5] performed reciprocating cylinder-on-plate tests with water lubrication. After running-in, the coefficient of friction increased from 0.03 to approximately 0.1. McKellop et al. [8] found dynamic coefficients between 0.07 and 0.12 which were slightly lower for cobalt-chromium than stainless steel. They used a pin-on-plate configuration and serum as lubricant. Wright *et al.* [9] found similar dynamic values (0.07...0.17) for different prosthetic materials using the same test configuration as McKellop. It is interesting to note that the initial static friction coefficient was higher with serum (0.11) than without (0.06). The variation of the reported values (0.04 to 0.2) has been explained by the differences between lubricants, sliding materials and sliding conditions [10]. In particular, the surface finish of the articulating bodies appears to have a major influence as shown by Fisher *et al.* [11] using a pin-on-disc apparatus: while the applied sliding velocity (35 re. 240 mm/s) had only a small effect on the determined friction coefficients, they did change with altered roughness of the stainless steel counterface from 0.07 to 0.2 and were found to be highest at a medium (!) roughness (range R_a: 0.014...0.078 µm).

In addition to the metal surface finish, the topography of the softer polymer may also influence the frictional properties of the articulation, as has been discussed theoretically by Moore [12]. This is of particular interest since the polyethylene surface of TKA plateaus undergoes changes in its characteristics during its time in situ: striated patterns with peaks and valleys are created, oriented along the principal directions of movement at the tibial articulation (section 6.3.3). It is hypothesized that this morphological change in the surface characteristics will alter the general frictional conditions under which the prosthesis is functioning and, thus, will modify its kinematic properties during its time in situ.

8.1.3 Purpose

The objective of this study was to evaluate the differences in friction between a smooth, freshly molded polyethylene surface and a worn surface, exhibiting a striated morphological change. In particular, the frictional characteristics should be analyzed with respect to varying slide-roll ratios of the femoral condyle on the tibial plateau to account for the specific contact kinematics in total knee arthroplasties.

8.2 Material and Methods

8.2.1 Friction Test Configuration

There are many methods which have been employed to measure the frictional force resulting from sliding motion under load. The magnitude of sliding friction can be easily monitored using a force or torque sensor in any continuous or reciprocating device. The investigation of friction during manifold slide-roll ratios, however, puts some constraints on the design of the testing apparatus. To achieve stable conditions, a reasonable length of the bearing surface has to be chosen. For instance the contact path during reciprocating motion as performed by the "Wheel-on-Flat" apparatus (chapter 7) is too short to acquire a stationary slide-roll ratio, suitable for precise friction measurements. Based on these considerations, a revolving simulator was developed capable of controlling the desired kinematics along an "infinite" course (Figure 8.2).

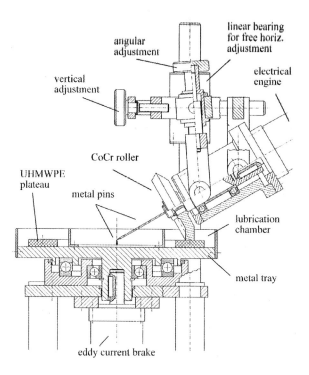

Figure 8.2: Mechanical configuration of the testing apparatus. During frictional measurements, the metal pins are replaced by flexible shafts, connected to an angular sensor. The normal force is applied to the free hanging unit comprising of vertical and angular adjustment and electrical engine (see Figure 8.5)

Kinematic Principles

As can be seen from Table 8.1, two principal kinematic modes are conceivable during rolling motion: (1) the velocity of the wheel is larger than the velocity of the plateau, resulting in an *accelerated* condition; or (2) vice versa, the wheel is slower than the plateau, generating a *decelerated* situation. Both modes can be conceptually modeled by either driving or braking the wheel on the plateau[*].

Test Concept and Mechanical Configuration

The interface of the revolving simulator is comprised by a cone shaped cobalt-chromium wheel (radius = 50mm) which is pressed onto a flat UHMWPE disc, generating a line contact. The CoCr roller runs along a 200 mm diameter course on the disc (Figure 8.2). The generated path corresponds to the curved path of the Miller-Galante knee system in the transverse plane, a result of the differences in radii between medial and lateral condyle (Figure 8.3). The UHMWPE disc itself is mounted on a horizontally and vertically pivoted metal tray to support the forces and moments transferred by the wheel. The wheel is pressed against the plateau using a pneumatic short stroke cylinder driven by a central pressure control device. The loading magnitude is adjustable and is recorded using a force sensor (range 0...2 kN; sensitivity ± 0.5%)[†].

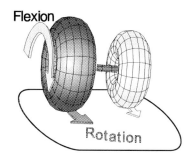

Figure 8.3: Knee flexion causes femoral rollback and rotation due to the bigger lateral condyle.

To exclude spin during the rolling motion of the wheel, its bearing surface is formed as a cone. A cylindrical wheel would generate varying surface velocities across the contact line and, thus, result in micro-slip. A cone shaped wheel, in contrast, progresses without superimposed spin because equal velocity profiles of disc and wheel can be achieved (Figure 8.4). In the accelerated mode, the roller is electromotively driven using an independent analogue loop to maintain the adjusted angular velocity, regardless the occurring loads at the driving shaft. The (vertically) free-hanging unit, consisting of roller and electric engine, can be adjusted manually in two directions to account for different geometries of the cone shaped roller. Metal pins (Figure 8.2) aid in defining the kinematically correct position.

---

[*] It should be noted that the accelerated mode corresponds to tractive forces (on the tibial plateau) against direction of movement, while the decelerated mode generates tractive forces in direction of motion. Thus, the accelerated mode matches the compressive and the decelerated mode suits the tensile protocol of chapter 7.

[†] Manufacturer addresses and technical data of the sensors are listed in appendix 8-I.

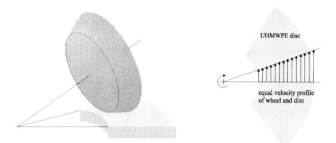

Figure 8.4: A cone-shaped wheel progresses without superimposed spin on the UHMWPE disc.

Control Mechanism and Data Acquisition

An eddy current brake serves to slow down the revolving metal tray. This type of brake has two advantages: it does not wear and it can be easily controlled by regulating the applied voltage. The control loop of the break is digitally established using a custom made program (source code Turbo Pascal®) on a PC*. An angular position sensor (continuous, sensitivity ±0.2°) is used to assess the angular velocity of the driven disc. It is fixed to the upper frame of the simulator and connected via a flexible shaft to the metal tray. The flexible shaft replaces the metal pin in Figure 8.2. The braking moment is adjusted to maintain a constant slide-roll ratio S:

$$S = \frac{\omega_d - 2\omega_w}{\omega_d} \cdot 100\% \qquad (8.1)$$

where ω_d is the angular velocity of the disc and ω_w the angular velocity of the wheel. By exchanging the actuators of the disc and the wheel, i.e. the eddy current brake is mounted on the wheel and the electric engine on the disc, the accelerated mode can be converted to the decelerated mode. The angular position sensor will then be connected to the wheel. A photograph of the set up of the friction apparatus is shown Figure 8.5.

The coefficient of friction for a defined slide-roll ratio $\mu(S)$ is calculated by the applied moment M_t of the brake and the normal force F_n

$$\mu(S) = \frac{M_t(S)}{l \cdot F_n} \qquad (8.2)$$

---

* Details are available from Reinholz [22].

where l is the effective lever-arm of the disc or the wheel (0.1m in the accelerated and 0.05m in the decelerated mode respectively). However, it did not seem appropriate to directly measure the moment M_t in this experimental set-up because the unknown friction of the ball bearing underneath the metal tray (Figure 8.2) would greatly affect the friction between wheel and disc. Therefore, the following calibration protocol was established which takes this into account.

Figure 8.5: Photograph of the testing device. The spring in the upper region of the simulator was replaced by a pneumatic short stroke cylinder to provide the normal load for this study. For a better overall view both kinematic modes are shown together. In the accelerated mode, the eddy current brake (left) is mounted underneath the metal tray, while the electric engine (right) is kept in place. It is vice versa for the decelerated mode. Here, the electric engine drives the metal tray, while the eddy current brake slows the CoCr roller.

Calibration Procedure

A torque sensor (20 Nm, ±0.5%) was set in between the driving shaft of the wheel and the vertical axis of the metal tray in order to bridge the interface of CoCr roller and polyethylene component (i.e. there was no connection between wheel and disc). Two uni-lever joints in series guaranteed proper alignment of the two axes and a constant angular velocity transfer. Using this set-up, the moment M_t (which would act directly at the interface between wheel and disc during testing) could be determined as a function of angular velocity ω and applied voltage U at the brake. This function includes the (unknown) frictional characteristics of the ball bearings[*] and, therefore, allows to determine the friction between wheel and disc without mathematical corrections after measurement.

$$M_t = f(U, \omega) \tag{8.3}$$

Measured moments were plotted against angular velocity ω_d (ω_w) at every 0.2 V for the accelerated (and decelerated) mode. The whole array of moments was fitted statistically with a multi variable non-linear regression approach (using SPSS 6.0 for Windows). A stepwise approach was selected to identify the variables from non-linear representations of U and ω which explained the highest amount of variance in the moments. These were then incorporated into the prediction equation for M_t and the coefficient of friction $\mu(S)$ was calculated for any given slide-roll ratio S during testing (according to equation 8.2).

8.2.2 Test Specimens

Two UHMWPE samples were tested: sample A ("new") was a direct compression molded disc with a surface morphology similar to those of unused Miller-Galante implants; sample B ("worn") was also compression molded, but with a machined surface finish to model the striated morphological change as found on retrieved tibial plateaus (Table 8.2). The cobalt-chromium wheel was similar to that of commercially available femoral condyles (R_a = 0.05 μm). An aqueous solution containing 30% bovine serum was filtered, stabilized and buffered to pH 7.6 to serve as lubricant (details see appendix 8-II).

Table 8.2: *Surface roughness of the polyethylene discs as compared to the roughness of the striated area of retrieved implants.*

UHMWPE	test sample A: "new"	test sample B: "worn"	implant retrieved[§]
R_{max} (μm)	3.56±0.37	6.86±0.20	6.77±2.41
R_z (μm)	2.56±0.45	6.51±0.20	5.57±2.41
R_a (μm)	0.41±0.10	1.57±0.10	1.24±0.45

[§]data from Table 6.3

---

[*]No normal load was applied on the ball bearings during the calibration procedure because their friction is hardly affected by normal load. Using an estimation according to Niemann [21] the torque increased by 0.043 Nm applying 1000 N normal load.

8.2.3 Test Protocol

The CoCr wheel was pressed with 1000 N onto the polyethylene surface. Measurements were taken at slide-roll ratios between 0 and 100% (10% steps) at four distinct, constant velocities (100, 140, 200, 280 mm/s). Data (U, ω, F_n) were recorded at each turn of the disc with at least 150 revolutions for every measurement step. Results were expressed as a mean ± (3× standard error + systematic error)[*] and plotted against the slide-roll ratio for every experiment (i.e. discrete mode, velocity, and sample). After applying the 1000 N normal force, each experiment started with 7800 revolutions at pure rolling and 1200 revolutions at a slide-roll ratio of 30% to assure proper running-in.

26 experiments were performed to investigate the influence of UHMWPE surface morphology, velocity and mode (acceleration / deceleration) on the frictional characteristics of femoral condyle and tibial plateau. To assure similar surface properties of the "worn" polyethylene disc throughout the study, a total of six samples was used. Every "worn" disc was replaced after three experiments, while eight experiments were conducted using the "new" disc. Oneway ANOVA, Tukey-B tests for multiple range analyses and Levene tests for homogeneity of variances were applied for statistical evaluation. The confidence interval was set to 95%.

8.3 Results

8.3.1 Moment/ Velocity Curves for System Calibration

A non-linear behaviour between angular velocity and moment was found for each single voltage applied (Figure 8.6). The characteristics of the moment curves M_t were similar, and a non-linear equation was defined to statistically describe the whole array for the accelerated and decelerated modes, respectively:

$$M_t = 1.4338 - 0.0066 \cdot \omega_d + 0.6293 \cdot \sqrt{\omega_d} \cdot U + 0.2736 \cdot \ln(\omega_d) \cdot U^2 - 0.9792 \cdot \sqrt[4]{\omega_d} \cdot U \quad (8.4a)$$

and

$$M_t = 0.9205 - 0.0050 \cdot \omega_w + 0.4907 \cdot \sqrt{\omega_w} \cdot U + 0.2905 \cdot \ln(\omega_w) \cdot U^2 - 0.6939 \cdot \sqrt[4]{\omega_w} \cdot U \quad (8.4b)$$

with U and ω explained in equation 8.3

---

[*] The equations for error calculation are plotted in appendix 8-III.

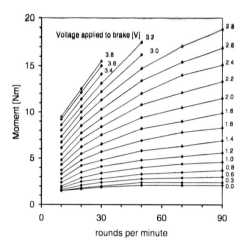

Figure 8.6: Moment/ velocity curves attained during calibration of the system

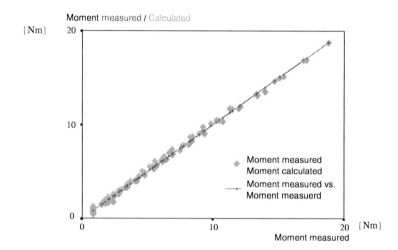

Figure 8.7: Relationship between calculated and measured moment for the accelerated mode. Similar results were obtained for the decelerated mode.

The relationship of the above theorems with the measured data is shown graphically in Figure 8.7. Multiple regression analyses confirmed that the theorems and the moment measurements correlated better than $r^2 = 0.926$ (mean (8.4a): $r^2 = 0.947$, mean (8.4b): $r^2 = 0.973$; p<0.05) for each distinct voltage value.

8.3.2 Influence of the Slide-roll Ratio, Kinematic Mode and Velocity

The maximum friction occurred neither with pure rolling nor with pure sliding, but with a combination of both. Hence, the coefficient of friction was affected by the applied slide-roll ratio. Plotting the friction coefficient versus increasing slide-roll ratio a curve with parabolic character (downside open) can be fitted. Representative examples for the "new" and the "worn" sample are shown in Figure 8.8.

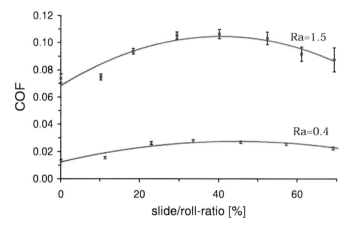

Figure 8.8: Typical plots (v = 140 mm/s; accelerated) of the coefficient of friction (COF) versus slide-roll ratio for a "new" ($R_a = 0.4$) and "worn" ($R_a = 1.5$) polyethylene surface; bars represent the measurement uncertainty according to appendix 8-III. Note that the friction coefficients are only plotted up to 70% slide-roll ratio because of technical problems encountered with the eddy current brake system.

While it was interesting to note that the "new" and the "worn" polyethylene sample responded in a similar manner to the applied slide-roll ratio, there was no significant influence of the mode (i.e. acceleration/ deceleration) and of the velocity on the coefficient of friction (in both cases p > 0.1).

8.3.3 Influence of Polyethylene Surface Morphology

The maximum coefficient of friction was significantly lower for the "new" disc at all tested slide-roll ratios independent of velocity ($p < 0.001$). The maximum coefficient of the smooth sample A was about three times smaller than the maximum coefficient of the rough, striated sample B (0.035 ± 0.007 vs. 0.102 ± 0.023). The relation between maximum coefficient of friction and the degree of applied sliding was similar for both surface morphologies ($54\pm14\%$ vs. $51\pm13\%$).

The friction coefficient applicable during pure rolling (slide-roll ratio = 0%) of the new polyethylene sample was significantly different from that of the worn polyethylene sample. While the coefficient of the smooth sample yielded rather low values of 0.023 ± 0.001, the coefficient of the rough sample yielded values of 0.061 ± 0.012. These values, however, were still significantly different ($p < 0.01$) from the maximum coefficient of friction which occurred at combined rolling and sliding.

8.4 Discussion

8.4.1 Surface Morphology Defines the Lubrication Regime

The results suggest that the morphological condition of the polyethylene surface is one of the most important factors for friction at the rolling-sliding contact as it occurs in total knee replacement. Furthermore, since the coefficient of friction was higher for the "worn" disc with the striated surface change it can be speculated that TKA dynamics will be altered with time in situ. If the coefficient of friction increases, tractive rolling of the femur on the tibia will become more likely and the articulation gets vulnerable to factors increasing the shear forces on the tibial plateau (section 5.4.2). This may have a detrimental effect on the wear of the plastic.

It is generally accepted that the magnitude of friction of components in a rolling-sliding contact is governed by the state of lubrication (section 3.6). As suggested by the friction values found in this study, a mixed lubrication regime was established at the interface between wheel and disc. For metal-on-plastic prostheses this has also been observed by others [13-15]. Further, it is in agreement with theoretical analyses applying elasto-hydrodynamic (EHD) theories: the lubricant film thickness is smaller than the surface roughness of the polyethylene articulation during gait [16]. Since the latter indicates local contact of the polyethylene asperities with the metal counterface, it is suggested that full fluid film formation is unlikely to occur during daily activities. Only under conditions of extremely light load, for example during the swing phase of gait, full fluid film formation takes place in the articulation as has been shown experimentally by Murakami [17].

No decrease in the friction coefficient was (statistically) determined with increasing velocity. At first glance, this is an unexpected finding because one would assume that friction depends on velocity in the mixed lubrication mode as suggested by the STRIBECK curve (Figure 3.17). Despite this theoretical mismatch, similar results have been published by Fisher at al. [11] who found that friction is independent of applied testing speed for a polyethylene pin in (purely) sliding contact with a stainless steel disc. While the aforementioned study was conducted at velocities of 35 and 240 mm/s to simulate the contact situation of the artificial hip, the applied velocities in this study ranged from 100 to 280 mm/s. It should be noted that *in vivo* the maximum speed at the hip's bearing surface is approximately 60 mm/s, whereas the maximum speed between femoral condyle and tibial plateau can reach 140 mm/s (as calculated from the data set of section 5.2). Thus, it can be speculated that the physiologically occurring velocity range is too small to show any *significant* effect[*] on friction when articulations with a low conformity are used. This is not the case for articulations exhibiting high conformity as demonstrated for the ball and socket joint [18].

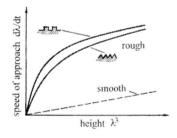

Figure 8.9: Speed of approach of two rectangular plates, whereby the lower surface exhibits a roughness profile. The difference in speed between a smooth and a rough surface becomes substantial, when the film thickness λ reaches values in the same order of magnitude as R_{max}. This effect is caused by the drainage of the fluid through the voids between the roughness peaks.

The striated morphological change on the surface did increase the coefficient of friction dramatically. Clearly, the striated pattern enhances the possibility of local contact between the first and second body. Under smooth surface conditions the lubricant can only exit through the circumference of the contact area, while the specific arrangement of the striations on the worn disc allows open channel flow. Thus, the fluid escapes through the valleys of the striated pattern and increases the number of contact spots between polyethylene summits and metal counterface. It is interesting to note that this effect is not only observable for EHD- lubrication, but also for squeeze film lubrication. Experiments, which have been conducted by Moore [12], demonstrate that the speed of approach of two rectangular plates is considerably faster when the topography of the surfaces in contact allows channel flow (Figure 8.9). Both, EHD- and squeeze film lubrication have been shown to be important lubrication mechanisms during dynamic loading of the artificial joint [15].

---

[*]for the given number of experiments in this study

8.4.2 Model Limitations

The limitations of the study shall be briefly discussed in this section. The specific design of the simulator required a mechanically stiff support of the revolving metal tray. In order to keep costs low, large ball bearings were applied instead of a thick-film hydrostatic support[*]. This design feature made it necessary to consider the frictional torque generated by the ball bearings. The placement of a moment sensor between the first and second body – the actual interface during testing – allowed to establish a calibration procedure which considered the torque of the ball bearings. Statistical analyses of the calibration data revealed that the moment acting at the interface (i.e. the frictional moment during testing) may be described as a non-linear function of the angular velocity and the applied voltage at the break. So, despite the use of ball bearings instead of hydrostatic bearings, the established protocol provided friction data unaffected by the mechanical set-up of the testing device.

The simulator produced unreliable motion characteristics when the slide-roll ratio was adjusted beyond 80%. The decreasing level of friction (see Figure 8.8) caused problems within the control loop in maintaining rolling at the articulation. Because the control mechanism's reaction was delayed, the metal tray stopped and disabled the brake system (an eddy current brake can only operate if there is motion). Therefore, friction was evaluated up to a slide-roll ratio of 70% where stable conditions were found throughout all experiments.

8.4.3 Friction in the Mixed Lubrication Regime

To the author's knowledge this is the first study which describes friction as a function of the effective slide-roll ratio for artificial knee implants. Maximum friction occurred at combined rolling and sliding and not – as it was assumed in chapter 5 – at pure rolling. The reason is evident: theoretical analyses were performed without taking the effects of elasticity as well as lubrication into consideration. Elasticity causes compliance between two bodies in contact and, hence, micro-slip during rolling motion at the interface. Pure rolling is therefore an ideal case which can never achieved in reality. Lubrication is also an important factor in rolling-sliding articulations as can be seen from Table 8.1. The occurring EHD film thickness is dependent on the *total velocity* of the first and second body. The total velocity reaches its maximum during pure rolling and has its minimum during pure sliding. In the rolling mode, both bodies

---

[*]hydrostatic bearings have been used by others to account for low friction of the testing device and, thus, accurate friction measurements of artificial joints [15]

transport lubricant and therefore increase the available fluid volume within the contact area. Thus, the separation of the two bodies is enhanced and friction is lowered during pure rolling.

It appears that two opposing friction mechanisms take place for the rolling-sliding contact in the mixed lubrication mode: increased static friction during pure rolling (effective for the intimate (dry) asperity contact), and increased film thickness during pure rolling (effective for the wet contact). This suggests that it is the interaction of both these mechanisms which is responsible for the maximum friction occurring at a slide-roll ratio of approximately 50%.

8.4.4 Effect of Friction Behaviour on TKA Kinematics

The maximum friction coefficients of the "new" and "worn" disc were approximately 60% higher than the coefficients during pure rolling. It was interesting to note that once the maximum was reached it dropped again considerably. Thus, if the tractive force on the tibial plateau exceeds a certain limit (e.g. because of too much acceleration or deceleration of the femoral condyle) the slide-roll ratio may take on values above 50% and friction between both contacting bodies will be reduced. Subsequently, posterior translation due to rolling of the femoral condyle on the tibial plateau will come to an end and pure sliding will occur. The femur will slide anteriorly or posteriorly with respect to the direction of the acting net shear force. These frictional characteristics of the rolling-sliding contact may explain the "jerky" discontinuous motion of knee implants recognized during fluoroscopic analyses [1,2,19] and simulation [3,4].

In conclusion, this study shows a complex relationship between slide-roll ratio and friction in total knee arthroplasties. The morphology of the polyethylene surface appears to have major influence on the generated friction magnitude. While there are other important factors, like counterface roughness [11] and normal load[*] [20] that should be considered, this study highlights that the development of a striated wear pattern on the tibial polyethylene plateau will increase friction at the articulation substantially. The relative importance of the coefficient of friction in generating tractive forces at the tibial plateau has been highlighted previously (e.g. section 5.4.1). Tractive forces damage the tibial plateau and lead to increased wear of the articulation.

---

[*] theoretical background see section 3.4.2

9 Wear Mechanism

9.1 Introduction

As reviewed in section 3.1.2, friction and wear are affected by the tribological system rather than by the intrinsic material properties of the damaged body. It is the input to the system, e.g. loading and kinematics, which ultimately determine the loss of material. In the previous chapters, several factors contributing to friction and wear of the tibial UHMWPE plateau in total knee arthroplasty (TKA) have been considered. The combined analysis of these results may provide an advanced understanding of the structure of the tibial tribosystem. This is important because a systematic improvement of the design and material characteristics of the tibial plateau requires not only knowledge about the effective wear mode but also about the acting wear mechanisms and their sequence in the history of the implant.

In chapter 5 tractive rolling was identified as the predominant dynamic mode occurring during gait in TKA. It was shown that tractive rolling increases the tangential forces at the tibio-femoral contact and surface fatigue occurs due to the specific contact movement. The magnitude of the tractive force is primarily influenced by the acting coefficient of friction. Due to rolling motion of the femoral condyles, a rolling abrasion type of wear is initiated when salt inclusions from the tibial polyethylene are detached (chapters 6 and 7). These salt inclusions are much harder than polyethylene and metal and, thus, act abrasive on the first and second body. The resulting wear pattern on the metal condyles consists of micron-sized scratches, aligned in the principal direction of knee motion (antero-posterior). The matching pattern on the tibial component is composed of micron-sized surface ripples, which are also A/P-oriented. It is interesting to note that these micro-features of wear at the tibial component are consistent with the morphological macro-change, consisting of A/P-oriented hills and valleys. This so-called striated pattern increases considerably the coefficient of friction at the wetted contact (chapter 8), which – in turn – increases the tractive forces.

Based on these observations, it appears that not a single wear mode and mechanism acts throughout the implant's life, but that modes and mechanisms vary with time and site. Before conclusions are drawn on the wear modes and mechanisms it is helpful to recall some important facts from the literature survey:

(1) The size of generated polyethylene particles in TKA barely exceeds 1-2µm (chapter 4.1.2, references 15-19).

(2) Engineering surfaces are rough and local contacts are established on asperity summits (chapter 3.3).

(3) Small particles (e.g. third bodies) have load carrying capacity [1].

(4) There is no *clinical* evidence for the belief[*] that higher tibial conformity reduces polyethylene damage (chapter 4.2.2).

In the past, most studies dealt with the *apparent* contact of the articulation rather than with the *real* contact established on asperities and third bodies. Using *global* stress analyses to study implant design and material issues, the facts 1 to 3 are not taken into consideration. While these global studies are essential to prevent total failure of the component, it needs to be acknowledged that polyethylene debris generation occurs at the micron-level as indicated by the produced particle size. Hence, it is suggested that *local* stresses, resulting from asperity and third-body contact at the polyethylene surface, determine volume, size and morphology of the generated debris. This seems to be reflected by a recent study from Bristol *et al.* [2] who found that there is no correlation between average contact stress and the amount of deleteriously loaded polyethylene contact area.

There are several studies pointing out that bone cement and metal particles are likely to be trapped inside the knee articulation (chapter 7.1.3). What is true for "external" particles resulting from the environment of the articulation, should be true also for "internal" particles detached from the articulating surfaces. The previously (section 6.3.5) identified salt minerals act as third bodies, once liberated from their polyethylene matrix (chapter 7). This is an important observation because third bodies do not behave statically but they participate in the kinematics of the articulation. Due to their potential load carrying capacity they may transform the entire kinetics of the contact. In other words: if the particles accumulate to a sufficient amount between the first and second body, the complete load transfer may be established on these particulates. Thus, it needs to be considered that the *real* area of contact is governed by third-body transport. This would be a feasible hypothesis to explain fact 4.

The observed striated morphological change at surface of polyethylene components (section 6.3.1) accounts for asperity contact between femoral condyle and tibial plateau. In this context two results need to be recalled: (a) it was shown that the striated pattern develops with implant duration, hence, wear mode and mechanism change with time; (b) the striated pattern occurs independent of design and, thus, may be an additional reason for the secondary influence of tibial conformity on the wear.

---

[*] established on stress analysis studies

In order to appropriately describe the wear history of the tibial component, it is reasonable to consider the *local* and the *global* scales of contact, each with its own set of damage mechanisms. In particular, the load carrying capacity of the previously identified salt particles is investigated, the resulting wear mechanism due to these third bodies is discussed, and their damage potential is evaluated in dependency of polyethylene oxidation. Further, based on the matching features between surface ripples and striations, a model for the development of the striated morphological change is presented. The significance of these striations on the wear history of the polyethylene component is then evaluated. Last but not least, the influence of the specific TKA contact kinematics, which have shown to accelerate fatigue, is disputed on a global basis.

9.2 Rolling Abrasion due to Salt Impurities

The analysis of retrieved tibial components (chapter 6) indicated the occurrence of salt inclusions within the polyethylene inserts. Once the machining marks are removed from the tibial component and the polyethylene surface becomes polished (burnished), these hard particles are found to be trapped inside the contact area after being detached from their matrix. They will be transported back and forth due to the reciprocating motion pattern of the knee and cause a *rolling abrasion* wear mode *("Kornwälzverschleiß")*. As known from other tribology studies (see [1]), third bodies can reside inside the articulation for a significant period of time until they are eliminated due to transportation towards the contact exits.

9.2.1 Contact Mechanics of Particle Indentation

The indentation of a plastic half-space by a rigid sphere has been extensively investigated, often in connection with indentation hardness testing. Hertz's analysis of the *elastic* stress field due to a spherical indentor on a flat surface shows that the maximum shear stress occurs at a depth of $0.53 \cdot a$ [3], where a is the radius of the contact circle and the material properties of polyethylene are applied (i.e. E = 572 MPa; v = 0.45). Hence, plastic yielding due to a salt particle would be expected to occur only slightly beneath the surface, since the diameter of the entrapped salt crystal is 10 microns at a maximum (Figure 9.1). These considerations are in agreement with the nano-fibrillar morphology found just beneath the surface of retrieved polyethylene specimens (chapter 6.3.6).

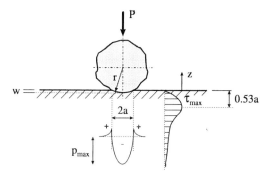

Figure 9.1: Schematic stress distribution for the elastic contact of a rigid sphere on a plane due to normal load*. Adapted from [9]

9.2.2 Load Carrying Capacity of the Salt Particles

Is the load distribution governed by the salt particles present? To answer this question, the contact problem sketched above (single indentor against plane) has to be transferred into a third-body problem. The third bodies can be handled as multiple, micro-indentors with roughly spherical surfaces (Figure 9.2).

From the Hertzian equations the partial load P on a single indentor on a flat plane (Figure 9.1) can be expressed as:

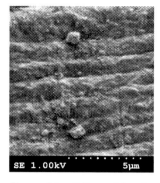

Figure 9.2: Salt particles in-between polyethylene ripples

$$P = \frac{4}{3} E' r^{\frac{1}{2}} w^{\frac{3}{2}} ; \qquad (9.1)$$

where E' is the elastic modulus according to equation 3.5, r the radius of the contacting indentor and w the compliance[†] of the bodies in contact. For multiple

---

[*] theoretical background see appendix 9.

[†] the distance which points outside the deforming zone move together

indentors (Figure 9.3), the latter can be expressed in terms of $w = z - s$, with z the height of the third-body particle, s the separation between the first and second body, and r_i the radius of the contacting area of the particle. Hence, the load P_i on any particle present is given by:

$$P_i = \frac{4}{3} E' \sqrt{r_i (z - s)^3} \; ; \qquad i = 1, 2, \ldots, n \tag{9.2}$$

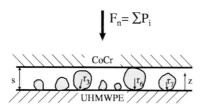

Figure 9.3: The first body (UHMWPE) and the second body (CoCr) are separated by third bodies (salt particles).

Following the model of Greenwood and Williamson [4] who presented a theory for asperity contact of rough surfaces, the probability of load support for any given particle, is the probability that its height z is greater than the separation s:

$$prob(z > s) = \int_s^\infty \phi(z) dz \; ; \tag{9.3}$$

where $\phi(z)$ is a probability density function describing the distribution of particle sizes. The expected number of third-body contacts will then be (if N is the total number of third bodies within the nominal contact area)

$$n = N \int_s^\infty \phi(z) dz \tag{9.4}$$

For simplification we assume now a complete spherical geometry of the particles. Then, the specific particle height z_i can be expressed in terms of $z_i = 2r_i$ and

$$P_i = \frac{4}{3} E' \sqrt{r_i (2r_i - s)^3} \; ; \tag{9.5}$$

As a first approach the particle radii r_i are assumed to be Gaussian distributed according to Figure 9.4. This assumption is based on the observation that most of the salt particles identified during retrieval analysis were 1 to 2 µm in size, with some up to 10 µm (chapter 6). It should be noticed that the assumption of a Gaussian distribution of irregularities on nominally flat surfaces has been shown to be a reasonable approach [4]. Introducing this Gaussian description for the particle distribution

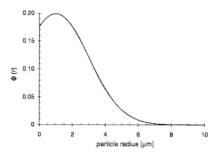

Figure 9.4: The assumed distribution of particle radii (based on retrieval findings of salt particles)

$$\phi(r) = \frac{1}{\sqrt{2\pi}\sigma} e^{-\frac{1}{2}\frac{(r-m)^2}{\sigma^2}} ; \qquad (9.6)$$

where m is the mean of the occurring radii and σ the standard deviation, the total load $F_n = \Sigma P_i$ carried by all n particles will be

$$F_n = \frac{4}{3} E'N \int_{\frac{s}{2}}^{\infty} \sqrt{r(2r-s)^3} \cdot \phi(r) dr ; \qquad (9.7)$$

To solve the above integral it is convenient to introduce standardized variables, and describe heights and radii in terms of the standard deviation σ of the particle distribution.

With $h = s/\sigma$, $\rho = r/\sigma$, and $\mu = m/\sigma$ follows:

$$F_n = \frac{2\sqrt{2}}{3\sqrt{\pi}} E'\sigma^2 N \int_{\frac{h}{2}}^{\infty} \sqrt{\rho(2\rho-h)^3} \cdot e^{-\frac{1}{2}(\rho-\mu)^2} d\rho \qquad (9.8)$$

The above equation can be solved using numerical integration methods. In Figure 9.5 – assuming the material properties of section 5.2.5 – the minimum force to achieve first and second body contact (i.e. $h = 0$) is plotted as a function of the total particle number N. Obviously, the more particles available the higher the load necessary to overcome the third-body separation. In other words: the load carrying capacity is linearly depending on the occurring particle amount N.

In chapter 5, using a mathematical model of the knee, the peak of the normal load was determined to 2.2 kN. According to the equation 9.8, approximately 2×10^5 particles have to accumulate to achieve a complete separation between tibial plateau and femoral condyles throughout the entire gait cycle (see Figure 9.5). Although this number may seem unrealistic, it should be noted that 2×10^5 particles of the above distribution cover less than 1% of the nominal (apparent) contact area[*] and represent the equivalent number of particles generated during one gait cycle in the hip [5]. Based on these considerations, a total separation of the knee articulation by particulates appears likely and provides an explanation that conformity is a secondary wear criterion.

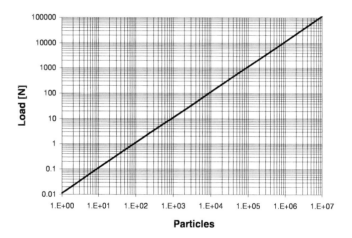

Figure 9.5: Calculated load carrying capacity of $1 \mu m$ spherical third bodies separating the (ideally smooth) surfaces of UHMWPE and CoCr. The table shows the load threshold until two-body contact occurs depending on particle number.

9.2.3 Wear Mechanisms due to Rolling Abrasion

In this section we assume a sufficient number of third bodies to allow full third-body contact. As the load increases, the size of each individual contact spot increases. However, since more third bodies come into contact the *mean* size of each third-body contact circle remains more or less constant. If the particles are swept back and forth

---

[*] based on the results of section 5.3.5 the apparent contact area is calculated to be 6mm × 20mm = 120 mm^2 (contact length for 2.2kN × condyle width). A densely packed mono-layer of 2×10^5 particles covers approx. 0.8 mm^2.

on the polyethylene surface, a relatively homogeneous contact pattern is generated, with parallel paths and a statistically uniform spatial distribution. This explains the regular spatial distribution of surface ripples observed on the retrieved polyethylene components (Figure 9.2). When the salt particles are detached from their matrix and free to move, *microploughing* is the predominant wear mechanism. It should be noted that the same wear mechanism occurs on the metal counterface, leaving A/P-oriented indentation marks at the surface. *Microcutting* occurs in combination with embedded, sharp edged particles (Figure 9.6). As illustrated in Figure 3.16, *rolling abrasion* under a *ploughing mechanism* causes mild wear on the tibial plateau. Under those

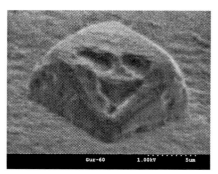

Figure 9.6: Salt crystal protruding from the UHMWPE surface

conditions, little if any material is removed from the surface; it is rather pushed along ahead and aside from the particle, in much the same way that a wrinkle can be formed with the finger tip dragged along a piece of cloth lying on the table.

9.2.4 The Influence of Polyethylene Oxidation

The plasticity index Ψ, given by equation 3.4, defines the onset of plastic flow. The latter helps to explain the observed differences in the rolling abrasion wear pattern between retrieved specimens and testing samples. While the testing samples were non-treated after fabrication, the retrieved specimens had been gamma sterilized at the time of manufacturing and, thus, had a different degree of oxidation (chapter 4.2.1, references 49-60). Although the *immediate* surface properties of gamma sterilized polyethylene components do not change dramatically with time in situ, density and crystallinity in the bulk of the component will increase considerably (Figure 9.7).

In other words: the hardness of the polyethylene sample is affected very little by oxidation, but the elastic properties are changed towards more brittleness. Using equation (3.4)

$$\Psi = \frac{E'}{H} \cdot \sqrt{\frac{\sigma}{r}}$$

Wear Mechanism 161

the plasticity index Ψ increases if the elastic modulus E' increases (and the hardness H = const.). Therefore, micro-ploughing and material flow is more likely to occur on (naturally) oxidized polyethylene implants than on non-oxidized samples.

a)
b)

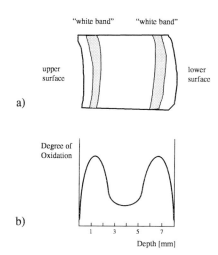

Figure 9.7: Cross sectional slice of an acetabular cup (a) illustrating the embrittled regions ("white bands") related to irradiation induced oxidation in UHMWPE (b). The interaction of gamma-rays with polyethylene causes free radicals which can not readily recombine and remain active for years. There are more radicals produced in the depth of the bulk. As oxygen diffuses in (preferred along the grain boundaries of polyethylene) chain scission and recrystallization occurs that embrittles the material. The diffusion of oxygen decreases with material depth. This explains the specific location of the embrittled regions. Adapted from [10, 11]

9.2.5 A Hypothesis for the Development of the Striated Pattern

As outlined in chapter 6.3.5 the salt crystals reside inside the polyethylene matrix in a cluster arrangement (Figure 9.8). Once liberated, the freely rolling particles form antero-posterior oriented "streets" (Figure 9.9) while more and more particles are released from the cluster reservoir with time. At locations of isolated A/P- knee motion, the width of each particle street is identical with the cluster diameter (Figure 9.9). At locations of superimposed spin, mostly in the posterior region of the implant, the particle streets are broader and interconnected (Figure 9.10). Thereby, the rolling particles generate surface ripples of a statistically uniform spatial distribution inside the street. Hence, more and more deformed material is pushed aside as

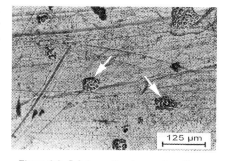

Figure 9.8: Salt impurities (arrows) on the non-worn surface of a direct molded component. The scratches on the polyethylene plateau are scratches replicated from the mold.

it would occur if the finger tip is drawn several times across the piece of cloth. It piles up and forms a hill when the sideways displaced material is stopped by an obstacle. In the case of the retrieved specimens the "obstacle" is the opposing flow of a neighbouring street (Figure 9.11): the material flow changes direction towards that of least resistance, namely the surface. With several salt clusters distributed across the tibial plateau, a striated landscape with numerous hills will be created (Figure 9.12). The height and shape of the summits will be influenced by the contact kinematics (→ particle transport) and the global stress and strain conditions.

 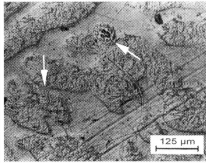

Figure 9.9: The impurities usually resided inside antero-posterior oriented streets exhibiting micro-ripples. The width of the street was often identical with the impurity diameter. Compare to Figure 6.15a.

Figure 9.10: Salt impurities (arrows) in the posterior region of the implant. Note the broader appearance and interconnection of the "particle streets".

UHMWPE

Figure 9.11: The freely rolling particles cause microploughing at the polyethylene surface and the material is pushed aside. The surface piles up and forms a hill if the flow of the opposing particle street meet each other.

Wear Mechanism 163

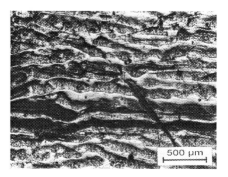

Figure 9.12: Striated landscape

9.2.6 Limitations of the Third-Body Damage Model

The calculations above provide further evidence that the salt inclusions in polyethylene are capable to completely separate the femoral condyles from the tibial plateau, thus, causing scratches on the metal component and material flow on the polyethylene insert. While these results suggest that stress analysis studies should be interpreted with care when predicting surface wear properties, the limitations of this study need to be noted. The Hertzian contact was applied to evaluate the influence of the salt particles on the polyethylene surface, hence it was assumed that the laws of continuum mechanics are still valid on the micron-level. Anisotropy and inhomogeneity of the polyethylene bulk were not considered and the elasticity properties of the metal component were taken on to be rigid. Also the characteristics of the salt particulates were eased assuming spherical and non-deforming third bodies. The minimum normal load to achieve contact between first and second body was calculated under the assumption of highly polished polyethylene and metal surfaces with negligible roughness. Furthermore, plasticity of the contacting bodies was not implemented into the model, although visibly present on the retrievals.

Nevertheless, the results appear to give an explanation to the observations made on retrievals and wear test specimens. The knowledge about the starting wear mode and mechanisms may provide the means to improve the bulk properties of polyethylene. In the next section, a subsequent wear mode – tractive rolling due to the striated wear pattern – will be illuminated with respect to the damage history and occurring wear mechanism.

9.3 Tractive Rolling due to Surface Striations

If it is assumed that the striated pattern has been created and there is a sufficient number of uniformly distributed hills to allow fully *macroscopic* asperity contact, then, the rolling abrasion wear mode will change to a two-body contact problem because most of the detached salt particles will drop into the surface valleys. The development of the striated pattern in the direction of motion reduces the effectiveness of the lubricant film due to "channel flow" (chapter 8.4.1) and, thus, increases the coefficient of friction. This mechanism is comparable to that of a profiled tire which helps to provide traction of the car on a wet street. On the polyethylene component, tractive rolling, inducing high shear forces at the contact region, becomes more likely under conditions of the striated surface pattern. In addition, the generation of residual subsurface strain and stress may be promoted due to asperity contact.

9.3.1 Wear Mechanism due to Tractive Rolling

Since direct contact spots between the metal and polymer bearing surfaces have been established, *adhesive* wear becomes predominant. In particular, polyethylene fibrils are generated (Figure 9.13) and ridges perpendicular to the direction of motion are established (Figure 6.18). As has been shown by wear tests (chapter 7), the latter are seen in the presence high tangential surface loads only (Figures 7.9, 7.10).

ripple pattern   summit   ripple pattern

Figure 9.13: Scan across the surface of a hill of the striated pattern using SEM. Note the polyethylene fibrils (arrows) on the summit.

Under lubricated conditions, sufficiently high tangential surface loads are produced in the presence of surface striations. As is shown in section 7.3.2 a tractive coefficient of at least 0.10 is needed to generate perpendicular ridges. This coefficient occurs under conditions of a striated morphological change of the surface. The ridges may initiate surface pealing with flaky debris.

9.3.2 Generation of Residual Strain and Stress

As discussed previously, the onset of plastic flow at asperities is mainly controlled by material and topographic properties and little by load (Figure 3.8). As proposed by the plasticity index Ψ (see above) high plastic strains are extremely likely to occur if the standard deviation of the asperity heights σ^* is increased. As reviewed in chapter 3.3.1 σ^* can be calculated from R_a for a Gaussian distribution of asperity heights:

$$\sigma^* = 1.25 \cdot R_a \tag{9.9}$$

Using the data of table 6.3 there is approximately a 10 -fold growth of σ^* when comparing the striated pattern to non-worn regions. Thus, Ψ increases more than 3 times and will be further increased due to changes in the elastic properties of the tibial components with time in situ. The deformation of the macroscopic polymer asperities within the striated pattern therefore manifests itself as residual strains below the surface (Figure 6.30). Their depth z can be estimated from the contact circle a and the material properties of polyethylene according to the Hertzian model [3]:

$$z = 0.53a ; \tag{9.10}$$

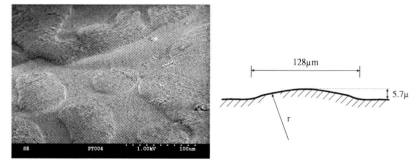

Figure 9.14: Protrusion of the summits within the striated pattern as seen by SEM imaging. The sketch shows the mean dimensions of a single summit.

The lateral width of the short pattern summits has been determined to have a mean of 128μm (chapter 6.3.3). As a first approximation, their shape can be considered spherical (Figure 9.13). Based on Figure 9.14, the radius of the asperity summits is estimated to be $r = 370\mu m$. Since the deformation of the asperities is not very sensitive to pressure (according to the theory of Greenwood and Williamson [4]), the two bodies in contact are separated by roughly the center line average of the asperities for a wide range of loads. Hence, the gap (separation) between femoral condyle and tibial plateau is approximately R_a and the compliance w of each asperity can be calculated knowing its height. Assuming that R_z defines the average height of all asperities,

$$w = R_z - R_a; \quad (9.11)$$

the contact radius a can be expressed in terms of

$$a = r^{\frac{1}{2}} w^{\frac{1}{2}}; \quad (9.12)$$

which gives

$$z = 0.53\sqrt{r(R_z - R_a)}; \quad (9.13)$$

Using $R_z = 5.7$ μm and $R_a = 1.24$ μm (table 6.3) the depth of maximum shear is calculated to be $z = 21\mu m$. This value coincides well with the experimentally determined depth of residual strains occurring beneath the summits of the striated pattern (section 6.3.6). It should be noted that the maximum shear stress moves closer towards the surface if tangential surface loads are applied. This explains the rather uniform distribution of residual strains versus depth in Figure 6.30.

9.3.3 Surface Fatigue due to Accumulated Stress

Under dynamic loading, localized strain and stress concentrations may have a profound effect on the damage of UHMWPE. Residual tensile stresses close to the surface may initiate cracks and accelerate their propagation. The origin of these cracks is supported by surface inclusions, i.e. occurring salt impurities. After crack initiation, the synovial fluid may play a crucial role in the pitting mechanism. Lubricant flows into the crack when it is passed by the femoral counterbody. According to a model of Way [6], the cyclic surface loading results in cyclic hydrostatic pressure in the fluid-filled crack, sufficient to cause extension of the crack and finally "pop-out" of the material.

Residual shear stresses below the surface may be responsible for sub-surface fatigue failure and delamination. Again, inclusions will work as crack initiators. The latter may propagate faster, if the UHMWPE powder particles are not sufficiently fused.

9.4 Wear Mechanisms due to Spin

All dynamic models presented in this thesis (chapters 5, 7, 8) have been established with the exclusion of spin of the femoral condyles on the tibial plateau. Although spin has been identified as an important kinematic mode (section 2.2.1), it has been neglected in an attempt to control as most variables as possible. From the retrieval analysis (chapter 6) as well as from the literature, however, it is suggested that polyethylene damage due to superposed spin should be considered.

Surface ripples with a similar appearance as the rolling abrasion pattern in the posterior portion (Figure 6.15b, *iii*) have been also found at the hip [7]. It was proposed that motion perpendicular to these ripples causes fibrillar pull-out due to *interfacial adhesion*. Most likely, this is also true for the knee at locations where rolling and sliding is superimposed by spin. These contact areas are then a source of considerable debris generation.

Abrasion may become an issue if spin of the knee causes the previously acquired antero-posterior directed scratches on the metal condyle to slide perpendicularly across the polyethylene surface. Recently, it has been shown by McNie et al. [8] that micrometer sized scratch lips are capable of producing a considerable amount of polyethylene wear. It is not the height of the scratch lip then which is the determining variable but its relative geometry "height/ half-width" (defined as aspect ratio). It was found that as the aspect ratio of the asperity lip increased, the plastic strains both on and below the surface of the UHMWPE increased non-linearly, offering the theory for the elevated wear rates seen with increased surface roughness (section 4.2.1, references 73-83).

9.5 Consequences

In this chapter it has been shown that total separation between the polished femoral condyle and the polished tibial plateau is likely to occur in the presence of detached salt particles. The load carrying capacity of these micron-sized third bodies has important implication for the wear history of the total knee joint. It is suggested that surface wear is driven by the *amount* of third bodies trapped within the articulation.

Hence, the time interval from particle entry to particle exit is crucial: the longer this time period the more particles accumulate. If the particle-threshold for separation is saturated, design considerations based on two-body contact are pushed into the background and *particle transport mechanisms* become effective.

While the above considerations are based on smooth surfaces of both the solid body and the counterbody, it has been outlined that macroscopic asperity contact – as it occurs in the presence of a striated pattern on the tibial surface – causes also localized stresses within the polyethylene. These may predominate stresses generated by the global (i.e. apparent) contact. Even more important, the occurring striations on the polyethylene component change the lubrication mechanism due to channel flow and permit higher tractive forces during rolling motion of the joint. A change in the coefficient of friction during normal function of TKA is generally not taken into account during the design process of the prosthesis. An increasing coefficient of friction, however, may have dramatic influence on the life-time shortage of the tibial plateau.

10 Summary and Conclusions

Total knee replacement has become a common surgical procedure, whereby problems due to wear are frequently reported. The purpose of the study was to investigate the tribological conditions at the tibial polyethylene component, in an attempt to better understand wear mode and mechanism. Loading and motion of the artificial knee were described, taking into account that friction is introduced by the utilization of metal-on-plastic bearings. The knowledge about the input parameters of the tribological system (i.e. loading and motion) permitted explicit analyses of friction and wear at the tibial polyethylene plateau. Subsequently, wear mode and lubrication regime were characterized applying contact conditions as they were identified on retrievals. It was found that polyethylene topography and the occurrence of intermediate material greatly affect friction and wear properties of the artificial joint.

Contact Mechanics at the Tibio-Femoral Joint

The normal knee typically articulates as a combination of rolling and sliding during flexion. This so-called femoral rollback is guided by the anterior and posterior cruciate ligaments. Although one or both of these ligaments are sacrificed during arthroplasty, friction at the artificial articulation still permits rolling motion during most of stance when the static coefficient of friction is equal to or above 0.2. However, this type of rolling, *tractive* rolling, introduces shear forces at the tibio-femoral bearing.

Using a mathematical model of the artificial knee, a peak tractive force of approximately 0.4 body weight was calculated, occurring coincidentally with a peak normal force of 3.3 body weight. Alterations in gait patterns had a substantial effect on tractive forces at the knee joint. When an abnormal gait pattern (often seen following total knee replacement) was input into the model, the posteriorly directed tractive force on the tibial plateau was reduced. It was also found that variations in muscle contractions associated with antagonistic muscle activity, as well as the angle of pull of the patellar ligament, affected the magnitude of tractive forces. Furthermore, the design of the polyethylene component influenced the contact forces. A slightly dished tibial design reduced the tractive forces compared to a flat design.

The specific characteristics of the tractive force, i.e. its alternating direction, plus the moving femoral contact generated cyclic stresses at the tibial plateau which provide conditions ripe for fatigue failure. The model exhibited that portions of the polyethylene surface were passed over three times during the stance phase of gait, and thus, were subjected to 5 Hz fatigue cycles during level walking. These accelerated fatigue conditions (3 million fatigue cycles per 1 million gait cycles) may be another factor (in addition to design) contributing to delamination and breakage of the polyethylene insert.

Early Wear at the Tibial Plateau

In addition to reported damage modes (e.g. pitting, polishing and scratching), subtle striations on the bearing surface of the tibial polyethylene plateau have been observed in components retrieved relatively early after implantation. The morphological characteristics of striations were consistent with the kinematic characteristics of the implant. The striated pattern was elongated and antero-posterior directed in the anterior bearing portion of the implant where rolling motion is expected. It was shorter and less organized in the posterior portion of the implant where rolling, sliding and spin act coincidentally. The topography of the pattern was consistent with its visual appearance, and the strong correlation between medial and lateral striated area suggested that this type of damage is initiated by cyclic rolling of the knee. Using a scanning electron microscope, the occurrence of micron-sized surface ripples inside the valleys of the striated pattern was shown. The ripple pattern showed nearly identical orientation and anisotropy as the striated pattern, indicating a direct link between micro- and macro-level of wear.

Low voltage imaging of the same, non-sputtered samples further revealed the occurrence of impurities at the surface as well as throughout the material. Their constituents were determined to be sodium, potassium and chlorine. Usually, the particles were 1 to 2 µm (occasionally 10 µm) in size, harder than the surrounding matrix, and occurred in clusters of 20 to 40 µm width. The salt inclusions were observed in all polyethylene retrievals, independent of design and manufacturer. In addition, they were identified in two freshly molded polyethylene sheets with and without calcium stearates. Hence, contamination due to *in vivo* body fluids or handling problems after retrieval appears implausible. In all likelihood the inclusions are a result of polymer synthesis, affecting most tibial liners currently available. Investigations of the nascent powder, as well as quantitative analyses of the salt content of different liners, remain to be done.

Initial Wear mode and prevalent mechanism

Using a wear testing apparatus, which accounted for the specific contact mechanics at the artificial knee joint, it was found that the micron-sized salt inclusions work as indentors. Once the particles are detached from their matrix, they cause a *rolling abrasion* type of wear with microploughing as the prevalent acting mechanism. A theoretical investigation provided evidence that these particles may accumulate to a quantity, sufficient to separate the femoral condyles from the tibial plateau.

This has several important implications for wear and wear analysis in total knee arthroplasty. If separation of the articulating surfaces takes place, the load transfer will be established on the present third bodies. Hence, contact between first and second bodies is lost and the design of the tibial polyethylene component (e.g. conformity) becomes a secondary wear criterion. This is reflected by clinical observation, since flat tibial components do not wear faster than dished ones. The results also suggest that stress analysis studies should be interpreted with care when predicting *surface wear* because most often an ideal geometry and perfect smoothness of the contacting bodies are assumed. Last but not least, the results point to the fact that the appropriate wear testing model has to be chosen when analyzing polyethylene wear. The time period that particles reside inside the articulation appears critical and depends on the contact kinematics of the joint. Continuous, uni-directional motion, as produced by a pin-on-disc machine, transports third bodies much faster out of the contact zone than reciprocating, multi-directional motion as present in the total knee. Thus, it is questionable if standard screening devices are appropriate predictors for polyethylene wear of the tibial component.

Change of Friction with Time in Situ

Using a revolving simulator, which was capable to control combined rolling and sliding, it was found that the morphological condition of the polyethylene surface is one of the most significant factors influencing friction. A striated morphological change at the tibial surface increases the coefficient of friction by approximately 300%, since the topography of the striated pattern allows channel flow of the lubricant. The striated pattern develops with time, making the coefficient of friction to a time dependent variable. Hence, the kinematics of the artificial knee will be altered, and rolling of the femur on the tibia becomes more likely with implant duration. While this is a desirable motion characteristic of the normal knee joint, the accompanying increased shear forces of tractive forces will increase wear at the polyethylene articulation.

In conclusion, wear at the prosthetic tibio-femoral joint is complex because the factors influencing the system are interdependent and time related. The salt inclusions of polyethylene appear to play a key role in the wear process and their function should be further investigated against salt-free polyethylene.

Appendix 5-I

Complete mechanical description of the knee model in Figure 5.3

The reference frame was fixed at the tibial plateau with its origin at the most anterior aspect of the polyethylene component. A left-handed coordinate system was used for the left knee, a right-handed for the right knee for description of the directions x, y, z:

x: lateral
y: anterior
z: proximal

$$\begin{bmatrix} M_x \\ M_y \\ M_z \end{bmatrix} = \begin{bmatrix} F_{Quad}[(c_i + R \cdot \alpha)\cos\beta + b \cdot \sin\beta] - F_{Hams}[(w_{ap} - c_i - R\alpha)\cos\alpha + a \cdot \sin\alpha] - F_{Gast}(w_{ap} - c_i - R \cdot \alpha)\cos\Theta + F_n \cdot f \frac{\omega}{|\omega|} \\ -\frac{1}{4}F_z w_{ml} - \frac{1}{4}F_{Tis,med} w_{ml} + \frac{1}{4}(F_{Quad}\cos\beta + F_{Hams}\cos\alpha + F_{Gast}\cos\Theta) w_{ml} + \frac{1}{2}F_{n,lat} w_{ml} + \frac{3}{4}F_{Tis,lat} w_{ml} \\ 0 \end{bmatrix}$$

$$\begin{bmatrix} F_x \\ F_y \\ F_z \end{bmatrix} = \begin{bmatrix} 0 \\ F_{Quad}\sin\beta - F_{Hams}\sin\alpha + F_{Gast}\sin\Theta + F_i \\ F_{Quad}\cos\beta + F_{Hams}\cos\alpha + F_{Gast}\cos\Theta + F_{Tis} + F_n \end{bmatrix}$$

with

$F_n = F_{n,med} + F_{n,lat}$

$\beta = \beta_0 - 0.47 \cdot \alpha$

$\beta_0 = 22° \dots 30°$

$\Theta = -3° + 0.75\alpha$

$\omega = \frac{d}{dt}\alpha$

and

M_x segmental flexion/ extension moment
M_y segmental abduction/ adduction moment
M_z segmental internal/ external moment (neglected)
F_x segmental external medial-lateral force (neglected)
F_y segmental external posterior-anterior force
F_z segmental external inferior-superior force

F_{Quad} force in the quadriceps group
F_{Hams} force in the hamstrings group
F_{Gast} force in the gastrocnemius group
F_{Tis} force in lateral or medial soft tissue
F_n normal force on the tibial plateau
F_t traction force on the tibial surface

α knee flexion angle
β patella ligament angle
Θ gastroc angle
ω angular velocity

a, b muscle insertion points (see figure 5.3b)
w_{ap} antero-posterior width of tibial plateau
w_{ml} medio-lateral width of tibial plateau
c_i initial contact point of the femoral condyles on the tibial plateau
R radius of the femoral condyles
f rolling lever (calculated according to appendix 5-II)

Appendix 5-II

Calculation of the rolling lever f (according to Johnson [19])

$f = 15 \times 10^{-4} \gamma \cdot a_c$;   free rolling

$f = \mu_t^2 \cdot a_c$;   tractive rolling

with

$$\gamma = \frac{1}{2}\left(\frac{[(1-2v_1)/G_1]-[(1-2v_2)/G_2]}{[(1-v_1)/G_1]+[(1-v_2)/G_2]}\right)$$

and

a_c semi-contact width
μ_t traction coefficient
v Poisson's ratio
G elastic shear modulus

Appendix 5-III

Gait data of two (representative) patients following total knee arthroplasty. The knee flexion angle is given in degrees, the forces are given in % Body Weight (BW), the moments are standardized in % BW × Height. (HWCF= Height weight correction factor)

AAZYWNL1	"Normal Pattern"		BW [N]	HWCF [Nm]
			635.6648	1049.483

angle (deg.)	Moment Mx	Moment My	Force Fy	Force Fz
-6.2719	-0.8351	-0.0243	6.9131	0.5628
-5.9781	-0.812	-0.0299	6.8143	0.7037
-5.3734	-0.7801	-0.0358	6.6676	0.918
-4.4814	-1.9379	0.1374	-3.0377	-21.9173
-3.3324	-2.8123	0.2869	-9.7134	-36.3241
-1.9637	-1.3816	0.3276	-7.7561	-51.426
-0.4187	-0.6679	0.2316	-7.6841	-67.5813
1.2539	-0.2743	0.1848	-8.1505	-82.9836
3.0016	0.3889	-0.1445	-6.1968	-95.3555
4.7706	1.0685	-0.6747	-3.3254	-103.3068
6.5087	1.5935	-1.3333	0.0518	-107.2691
8.1673	2.26	-1.9039	4.1927	-108.9273
9.704	2.9875	-2.8605	9.1257	-110.1058
11.0842	3.5425	-3.4924	13.1763	-110.3326
12.2817	3.5986	-3.793	15.5391	-108.7975
13.2792	3.6449	-4.0441	17.3758	-108.3973
14.0674	3.7784	-4.0045	19.5859	-107.5792
14.6446	3.6837	-3.7799	20.5557	-105.7075
15.0153	3.4216	-3.5973	20.7824	-103.0697
15.1894	3.2403	-3.3517	20.8918	-99.4011
15.1811	2.9436	-3.1897	20.5332	-94.9558
15.0076	2.7312	-2.933	20.2501	-90.331

14.6891	2.4467	-2.731	19.8136	-85.3227
14.2471	2.1809	-2.5206	19.1423	-81.2806
13.7046	1.8339	-2.3827	18.4051	-78.1842
13.0851	1.5832	-2.2794	17.8249	-75.5369
12.412	1.2026	-2.2249	17.0127	-72.9425
11.7086	0.9803	-2.1763	16.4086	-70.1
10.9972	0.7201	-2.177	15.9161	-67.7027
10.2985	0.4742	-2.1405	15.3502	-65.7683
9.6318	0.2069	-2.1183	14.8309	-64.2833
9.0135	0.1341	-2.0798	14.7213	-63.172
8.4578	-0.0352	-2.0949	14.8385	-62.6977
7.9759	-0.1485	-2.1045	14.7121	-62.0406
7.576	-0.3674	-2.1376	14.5174	-62.0876
7.2633	-0.5439	-2.1283	14.3436	-63.0467
7.0406	-0.6817	-2.1291	14.5902	-64.6063
6.9083	-0.8212	-2.0779	14.6488	-66.4523
6.8653	-0.9075	-2.1086	15.1254	-68.6797
6.9096	-0.9577	-2.0727	15.727	-71.3481
7.039	-1.0081	-2.2097	16.6248	-74.2048
7.2516	-0.9767	-2.2737	17.908	-77.4227
7.5467	-0.9652	-2.3127	19.3081	-80.8609
7.9251	-0.843	-2.3093	20.8699	-84.5002
8.3894	-0.8206	-2.395	22.5759	-88.6024
8.944	-0.6101	-2.3995	24.5272	-92.6439
9.5954	-0.4604	-2.5227	26.1528	-96.1456
10.3518	-0.2601	-2.5801	27.8287	-99.6768
11.2228	-0.034	-2.7023	29.2809	-102.3991
12.219	0.3376	-2.7417	30.7885	-104.6554
13.3512	0.6377	-2.7727	31.6964	-105.5305
14.6303	1.0181	-2.7651	32.3762	-105.5866
16.0663	1.0927	-2.7996	31.6546	-104.7462
17.6673	1.5716	-2.7617	31.6613	-102.7755
19.4391	1.7965	-2.7118	30.6322	-99.4024
21.3842	2.1158	-2.5147	29.3086	-94.9195

23.501	2.2541	-2.3819	27.1994	-89.106
25.7829	2.2383	-2.028	23.7542	-81.1875
28.2181	2.0778	-1.6767	19.3794	-71.0453
30.7891	1.9687	-1.0615	14.6081	-57.9835
33.4726	1.8241	-0.5993	9.9811	-43.7023
36.24	2.1351	-0.3106	7.3006	-30.9572
39.0583	2.1517	-0.1511	4.5938	-21.5801
41.891	1.8683	-0.0168	1.569	-14.8704
44.7	1.6318	-0.0137	-0.54	-9.9082

AAYVWSR0	"Quadavoidance"	BW(N)	HWCF(Nm)
		686.7847	1194.937

angle (deg.)	Moment Mx	Moment My	Force Fy	Force Fz
-6.0205	-0.2814	-0.1944	3.558	3.4206
-5.7662	-0.2748	-0.2002	3.4329	3.4831
-5.3856	-0.2686	-0.1998	3.2986	3.538
-4.8939	-2.4053	0.5756	-6.3111	-12.909
-4.3073	-1.9621	0.6626	-5.5644	-16.693
-3.6432	-1.2602	0.5179	-3.6911	-21.0484
-2.9195	-1.4301	0.686	-5.2944	-28.0745
-2.1539	-1.6783	0.8513	-6.9592	-35.5013
-1.3637	-1.5343	0.9345	-6.9829	-42.3166
-0.5655	-1.4663	1.1633	-6.9987	-50.1297
0.2257	-1.3623	1.113	-6.6869	-56.9223
0.9964	-1.3916	0.8352	-6.5927	-61.5172
1.7349	-1.4215	0.0515	-6.4092	-63.8199
2.4313	-1.3885	-1.0349	-5.9775	-65.133
3.0778	-1.4388	-1.8962	-5.6629	-65.9744
3.6684	-1.6431	-2.1773	-5.4893	-68.4193

4.1989	-1.3795	-2.3252	-3.6727	-72.1034
4.6666	-1.1485	-2.1627	-1.9764	-74.6469
5.0705	-1.3803	-1.9428	-1.7802	-76.7723
5.4104	-1.268	-2.0671	-0.4926	-79.4346
5.6875	-1.3077	-2.0891	0.373	-82.2301
5.9039	-1.2482	-2.0944	1.4577	-85.0273
6.0623	-1.2795	-2.4031	2.3399	-87.6113
6.1658	-1.2509	-2.7844	3.1395	-89.7791
6.2182	-1.2027	-3.3109	4.0425	-91.336
6.2235	-1.153	-3.7761	4.9186	-93.0963
6.1859	-1.1762	-4.0024	5.4739	-94.4216
6.1095	-1.1656	-4.0999	6.0123	-95.9544
5.9988	-1.1264	-4.1905	6.6322	-96.0554
5.8581	-1.0957	-4.1343	7.2197	-95.9581
5.6915	-1.0683	-4.0187	7.5617	-95.0471
5.5031	-1.0857	-3.9506	7.7515	-93.7484
5.297	-1.1071	-4.0245	7.8824	-91.8599
5.0769	-1.103	-4.0893	8.0879	-90.196
4.8467	-1.1483	-4.2487	7.9327	-88.152
4.6097	-1.139	-4.2705	7.9534	-86.1124
4.3695	-1.1766	-4.3951	7.927	-83.8876
4.1293	-1.1741	-4.4036	7.8066	-82.085
3.8922	-1.2203	-4.3147	7.7908	-80.6907
3.6615	-1.1716	-4.2	8.0966	-79.4865
3.4399	-1.2001	-4.1512	8.3228	-78.6997
3.2303	-1.1277	-4.1105	8.8878	-78.0917
3.0351	-1.158	-3.9666	9.2959	-77.9002
2.8569	-1.1402	-3.8292	9.9749	-78.0861
2.6978	-1.1884	-3.8886	10.3465	-78.097
2.5594	-1.187	-3.7059	10.8774	-78.2741
2.4435	-1.2995	-3.6636	11.1495	-78.6835
2.351	-1.2997	-3.5791	11.7985	-79.2398
2.2827	-1.2803	-3.5637	12.5141	-79.9795
2.2389	-1.3377	-3.4844	13.0958	-80.9303

2.2193	-1.4194	-3.4629	13.6171	-81.8833
2.2234	-1.4104	-3.4074	14.4231	-82.9869
2.2503	-1.4271	-3.487	15.187	-83.6718
2.2985	-1.4185	-3.5068	16.0065	-84.96
2.3665	-1.4161	-3.4575	16.822	-86.0282
2.4524	-1.3879	-3.4565	17.7177	-87.2892
2.5541	-1.3491	-3.5063	18.5507	-88.1385
2.6699	-1.3197	-3.5717	19.3541	-88.7806
2.7979	-1.3319	-3.5671	20.2182	-89.6192
2.9364	-1.2885	-3.5562	21.1545	-90.6548
3.0843	-1.3291	-3.6429	21.792	-91.3509
3.241	-1.2491	-3.6108	22.7947	-92.1647
3.4067	-1.293	-3.6997	23.2902	-92.7116
3.5821	-1.2475	-3.6449	23.9506	-93.433
3.7693	-1.2599	-3.7107	24.4833	-93.7857
3.9709	-1.1906	-3.7456	24.9612	-93.958
4.191	-1.2169	-3.8374	25.3004	-94.4172
4.4346	-1.1574	-3.8749	25.6578	-94.8947
4.7079	-1.1458	-3.8943	25.9501	-95.1961
5.0179	-1.0292	-4.0803	26.534	-95.6453
5.3728	-0.9101	-4.0141	26.99	-95.7045
5.7814	-0.7424	-3.9496	27.4622	-95.9959
6.2531	-0.6709	-3.9521	27.623	-95.9958
6.798	-0.5034	-3.9139	27.8544	-95.5331
7.426	-0.3428	-3.6669	28.1295	-95.0575
8.1472	-0.1845	-3.4894	28.0074	-94.0986
8.9712	0.0018	-3.2459	27.9322	-92.442
9.9072	0.1695	-2.9594	27.7055	-90.4054
10.9632	0.3765	-2.7109	27.0122	-87.2171
12.1464	0.5412	-2.2927	26.0064	-83.4806
13.4623	0.7417	-1.9629	25.0261	-79.2413
14.9147	0.8972	-1.5184	23.6442	-74.2441
16.5055	1.0539	-1.1114	22.011	-68.6382
18.2341	1.1932	-0.5764	20.1553	-62.3745

20.0976	1.2393	-0.1867	17.69	-55.2005
22.0901	1.3596	0.2909	15.0788	-47.005
24.203	1.4883	0.6051	12.5168	-38.3662
26.4248	1.6158	0.9096	10.0746	-30.2334
28.7407	1.6677	0.9357	7.5534	-22.5215
31.1332	1.5428	0.7613	4.9103	-16.3932
33.5819	1.4269	0.5665	2.6873	-11.126
36.0645	1.443	0.4052	1.3083	-7.4794
38.5561	1.5326	0.2848	0.7121	-4.6714

Appendix 5-IV

Gait analysis – Methodology and system set-up

Gait analysis data were obtained based on the idealization that considered the lower extremities as a rigid-body linkage with movable joints at the ankle, knee, and the hip. In this analysis the joints were assumed to have fixed axes of motion in each plane (similarly to hinge joints). Further it was assumed that the flexion-extension axis remained perpendicular to the plane of progression and that the axes for abduction-adduction and internal-external rotation moved with the body segments. Inertial properties of foot, shank and thigh were implemented to account for angular velocity and acceleration about the perpendicular axes of the limb segments.

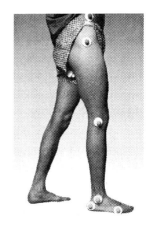

Figure: Application of reflective diodes for gait analysis

Three-dimensional motion of the lower limbs was assessed using a two-camera optoelectronic digitizer and six light-emitting diodes which were placed at the anterior superior iliac spinae, at the center of the greater trochanter, over the mid-point of the lateral joint line of the knee, on the lateral aspect of the malleolus at the ankle, at the base on the calcaneus, and the base of the fifth metatarsal (see Figure). The location of the joint centers at the hip, knee and ankle was estimated relatively to the position of the markers according to a protocol as described by [59].

The position of each marker was sampled seventy-five times per second and tracked using an electronic signal-conditioning device. The vector of the ground reaction force (and any occurring twisting moment) was recorded simultaneously using a piezoelectric force platform. The calculated external forces and moments of the body segments at each joint were plotted relative to the anatomical axes of the segments, e.g. the flexion-extension moment of the knee was plotted relative to a coordinate system fixed in the shank. EMG data were acquired using surface electrodes. Data were triggered to heel-strike and toe-off, while the threshold "on-off" was determined according to [60].

Appendix 5-V

ADINA contact algorithm

The majority of commercially available finite element codes use a contact algorithm which employs the use of so-called *gap* elements. These elements occupy the empty space between the two contacting bodies and close the contacting bodies near each other. Once in contact, the gap elements provide the necessary stiffness to prevent overlap of the contacting bodies. Thus, the contact is non-linear due to the dependence of the stiffness of the gap elements on the relative displacement of the two contacting bodies.

In ADINA, a contactor node surface and target surface are specified. The algorithm criterion is to determine the degree of material overlap which occurs at each equilibrium iteration. The algorithm consists of the following four steps: first, determine all possible target surfaces for each contactor node; second, find the target node closest to each contactor node; third, determine the degree of material overlap using the local nodal coordinates; and fourth, obtain the dot product (*"Skalarprodukt"*) of the normal vectors of the contactor and target surface. Contact is admissible only, if the dot product is a negative value [33]. Thus, the contact algorithm in ADINA is not dependent in the discretization of the gap elements, but the actual model itself. Similarly, the conditions for closure of the space between the two contacting bodies is dependent on the stiffness of the actual contacting bodies as opposed to the stiffness of the gap element.

Appendix 6

Retrieval Data

Access No.	Manufacturer	Design	Time in situ (mo)	Age (years)	Sex	Side	Reason for Removal
001	Zimmer	Miller-Galante	37	55	female	right	pain
002	Zimmer	Miller-Galante	8	68	male	left	instability
003	Zimmer	Miller-Galante	23	61	male	right	patella subluxation
004	Zimmer	Miller-Galante	13	78	female	left	instability
007	Zimmer	Miller-Galante	40	71	female	left	pain
174	Zimmer	Miller-Galante	19	31	female	right	pain
229	Zimmer	Miller-Galante	2.5	66	female	*	infection
236	Zimmer	Miller-Galante	5.5	65	male	right	infection
237	Zimmer	Miller-Galante	15	62	female	right	pain
282	Zimmer	Miller-Galante	2.5	64	female	left	malrotation
394	Zimmer	Miller-Galante	23	61	male	right	infection
S96-14630	Zimmer	Miller-Galante	143	58	female	left	broken femur above TKA
S96-13801	Zimmer	Miller-Galante	103	72	male	left	*
S96-08379	Zimmer	Miller-Galante	92	75	female	left	instability
S96-11513	Zimmer	Miller-Galante	103	70	female	left	*
S96-03826	Zimmer	Miller-Galante	104	76	female	right	*
003b	Zimmer	MG II	17	65	male	left	patella subluxation
009	Zimmer	MG II	0.3	62	female	left	dislocated patella
369	Zimmer	MG II	2	*	*	*	*
385	Zimmer	MG II	8	64	female	right	loosening
447	Zimmer	MG II	26	72	female	right	autopsy

S96-13577	Zimmer	MG II	36	70	female	left	patellar maltracking
S96-13739	Zimmer	MG II	7	75	female	right	*
S96-10157	Zimmer	MG II	65	59	female	left	infection
S96-XXXX	Zimmer	MG II	29	47	female	right	infection
S96-01036	Zimmer	MG II	30	72	female	right	hemorhagic synovitis
S96-00473	Zimmer	MG II	29	78	male	right	instability
MM.6.26.91	Zimmer	IB-II	27	77	female	right	pain, instability
MB.7.21.93	Zimmer	IB-II	28	75	female	right	loosening
JT.1.12.94	Zimmer	IB-II	6	60	female	left	infection
AA.5.18.94	Zimmer	IB-II	5	77	female	left	failed MCL
VP.7.25.94	Zimmer	IB-II	29	79	female	right	pain, loosening
SV.6.28.93	Zimmer	IB-II	84	55	female	*	*
S96-04282	Zimmer	IB-II	41	52	female	right	instability
S96-05243	Zimmer	IB-II	8	72	female	left	pain
S96-03505	Zimmer	IB-II	1	78	male		
S96-09095	Zimmer	IB-II	12	71	male	right	pain, stiffness, tibial malalignment
S96-14098	Zimmer	IB-II	30	76	female	left	pain
S96-YYYY	Zimmer	IB-II	24	52	female	left	infection
S97-14124	Biomet	AGC	6	68	male	right	pain; loose tibial component
DD 1485-52	DePuy	AMK CR	*	*	*	*	*
DD 1486-50	DePuy	AMK PS	*	*	*	*	*
S97-08750	Howmedica	Duracon	14	44	male	right	loose tibial component
S96-16564	Johnson &Johnson	*	93	66	female	right	infection
K3073/95L# 7	Osteonics	7000 Series	28	69	female	left	autopsy
S97-13285	Smith Nephew	Genesis II	4.5	61	male	left	instability
S97-02185	Wright	IB2 PS	17	65	male	left	infection

* information not available

Appendix 7-I

Description of sensors

Force transducers

Description, Type	Force transducer with gage measurement system, C9B	Load cell, U 2A
Mode	Compression	Tension/Compression
Company	HBM (Hottinger Baldwin Messtechnik)	HBM (Hottinger Baldwin Messtechnik)
Location	Darmstadt, Germany	Darmstadt, Germany
Nominal force	2 kN	±500 N
Nominal value	1 mV/V	2 mV/V
Deviation from nominal value	≤ 1 %	Tension: ≤ 0,2 % Compression: ≤ 1,5 %
Non-linearity	≤ ± 0,5 %	≤ ± 0,2 %

Displacement transducers

Description, Type	Potentiometric displacement transducer, T 100	Potentiometric angular displacement transducer, 8820
Company	Novotechnik	Burster
Location	Ostfildern (Ruit), Germany	Gernsbach, Germany
Measurement range	106 mm	350° ± 2°
Non-linearity	± 0,075 %	0,5 %
Service life	> 100 × 10^6 strokes	> 20 × 10^6 revolutions

Appendix 7-II

Pneumatic Set-up of the "Wheel-on-Flat" Apparatus

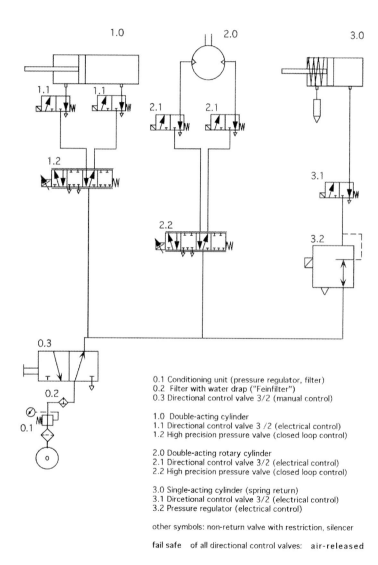

0.1 Conditioning unit (pressure regulator, filter)
0.2 Filter with water drap ("Feinfilter")
0.3 Directional control valve 3/2 (manual control)

1.0 Double-acting cylinder
1.1 Directional control valve 3/2 (electrical control)
1.2 High precision pressure valve (closed loop control)

2.0 Double-acting rotary cylinder
2.1 Directional control valve 3/2 (electrical control)
2.2 High precision pressure valve (closed loop control)

3.0 Single-acting cylinder (spring return)
3.1 Dircetional control valve 3/2 (electrical control)
3.2 Pressure regulator (electrical control)

other symbols: non-return valve with restriction, silencer

fail safe of all directional control valves: air-released

Appendix 8-I

Description of sensors

Force / torque transducers

Description, Type	Ultraminiature force sensor, 8416	Torque sensor, TD020G
Mode	Compression	-
Company	Burster GmbH	ASM GmbH
Location	Gernsbach, Germany	Unterhaching, Germany
Range	0 ... 2 kN	0 ... 20 Nm
Nominal value	1 mV/V	-
Deviation from nominal value	≤ 0,5 %	-
Non-linearity	-	≤ ± 1 % FS

Displacement transducer

Description, Type	Analogue angular displacement transducer, AWS 1
Company	ASM GmbH
Location	Unterhaching, Germany
Measurement range	350°
Non-linearity	0,10 % / 0,5 %
Service life	> 100 x 10^6 revolutions

Appendix 8-II

Lubricant for Friction Analyses

Materials and Apparatus

- sodium chloride
- 1-phenoxy-2-propanol ($C_9H_{12}O_2$)
- Tris-hydroxy-methylamine ($C_4H_{11}NO_3$)
- concentrated Hydrochloric acid (HCl_{conc})
- Bovine serum
- Deionized or distilled water
- Apparatus for membrane filtration
- Membrane filters of pore sizes 0.8 µm and 0.2 µm

Preparation of lubricant

For 300 ml solution:

Prepare 130 ml solution containing 9g/L of sodium chloride in distilled water. Heat the solution to approx. 50°C and add 3 ml $C_9H_{12}O_2$. Allow the solution to cool and add 5.4g $C_4H_{11}NO_3$. When the crystals have dissolved, adjust the pH to 7.6 by adding a v/v 50% aqueous solution of HCl_{conc}. Add distilled water up to a total volume of 200 ml, then mix it with 100 ml bovine calf serum.

Filter the mixture through a membrane filter of pore size 0.8 µm, and filter again directly into the chamber of the test apparatus through a membrane filter of 0.2 µm pore size.

Appendix 8-III

Error determination

In this study, the uncertainty of the measurement was expressed as weighted sum of statistic and systematic errors. It may be shown statistically that if the number of recorded data exceeds 100, the real value of a single measurement point will be within the range of mean ± 3·standard deviation for about nearly 100%[1]. Then, the uncertainty Δ of the measurement can be described as

$$\Delta = |3\sigma| + |s_e|$$

where σ is the standard deviation and s_e is the systematic error of the measurement device (e.g. due to non-linearity).

The coefficient of friction μ was determined as a function of applied voltage U, angular velocity ω, and normal Force F_n

$$\mu = f(U, \omega, F_n)$$

Then, the standard deviation σ_μ of μ can be calculated according to

$$\sigma_\mu = \sqrt{\left(\frac{\partial \mu}{\partial U}\right)^2 \cdot \sigma_U^2 + \left(\frac{\partial \mu}{\partial \omega}\right)^2 \cdot \sigma_\omega^2 + \left(\frac{\partial \mu}{\partial F_n}\right)^2 \cdot \sigma_{F_n}^2}$$

---

[1] Dehoust O. *Physikalische Praktikum*, München: TU-Verlag, 1986

Appendix 9

The Hertzian theory is often chosen as a first approach to predict contact stress levels. However, very often the assumptions behind this theory are not clearly defined, therefore they are re-stated as follows:

(1) the surfaces in contact are frictionless

(2) the surfaces are continuous and non-conforming

(3) each solid can be considered as an elastic half-space

(4) the strains are small: $a \ll R$ (with a = radius of the contact circle, and R the relative curvature of the two solids: $1/R = 1/R_1 + 1/R_2$)

The maximum Hertzian pressure can then be calculated according to:

$$p_{max} = \sqrt[3]{\frac{6F_n \cdot E'^2}{\pi^3 R^2}} \; ;$$

and the radius of the contact circle:

$$a = \sqrt[3]{\frac{3F_n \cdot R}{4E'}} \; ;$$

where F_n is the normal load and E' the relative Young's modulus according to equation 3.5.

The stresses along the z-axis may be calculated using:

$$\frac{\sigma_r}{p_{max}} = \frac{\sigma_\theta}{p_{max}} = -(1+v)\left[1-(z/a)\tan^{-1}(a/z)\right] + \tfrac{1}{2}(1+z^2/a^2)^{-1} \; ;$$

$$\frac{\sigma_z}{p_{max}} = -(1+z^2/a^2)^{-1} \; ;$$

whereby the principal shear stress is defined according to $\quad \tau_1 = \tfrac{1}{2}|\sigma_r - \sigma_\theta|$.

References

Chapter 1

1. Knutson, K., Lewold, S., Robertsson, O., and Lidgren, L. The Swedish knee arthroplasty register: A nation wide study of 30,003 knees 1976-1992. *Acta Orthop Scand* **65**:375, 1994.

2. Rand, J. A. and Ilstrup, D. M. Survivorship analysis of total knee arthroplasty. *J. Bone Joint Surg* **73-A**:397, 1991.

3. Windsor, R. E., Scuderi, G. R., Moran, M. C., and Insall, J. N. Mechanisms of failure of the femoral and tibial components in total knee arthroplasty. *Clin. Orthop.* **248**:15, 1989.

4. Scott, W. N. *The Knee*, St. Louis: Mosby, 1994.

5. Gluck, T. Die Invaginationsmethode der Osteo- und Arthroplastik. *Berl. klin. Wschr.* **27**:732, 1890.

6. Scott, R. D., Joyce, M. J., Ewald, F. C., and Thomas, W. H. McKeever metallic hemiarthroplasty of the knee in unicompartmental arthritis. *J. Bone Joint Surg* **67-A**:203, 1985.

7. Rand, J. A. Introduction. In J.A. Rand (Ed.), *Total Knee Arthroplasty*. New York: Raven Press Ltd., 1993.

8. Gunston, F.H. Polycentric knee arthroplasty. Prosthetic simulation of normal knee movement. *J Bone and Joint Surg* **53-B**:272, 1971.

9. Conventry, M. B., Jackson, E. U., Riley, L. H., Fimerman, G. A. M., and Turner, R. H. Geometric total knee arthroplasty; I. conception, design, indications, and surgical technic. *Clin Orthop* **94**:171, 1973.

10. Insall J., Ranawat C. S., Scott W.N.S., and Waler P. Total condylar knee replacement: preliminary report. *Clin Orthop* **120**:149, 1976.

11. Robertsson, O., Knutson, K., Lewold, S., Goodman, S., and Lidgren, L. Knee arthroplasty in rheumatoid arthritis. *Acta Orthop Scand* **6**:545, 1997.

12. Hip and knee implant review. *Orthopedic Network News* **6**:1, 1995.

13. Dorr, L. D. and Serocki, J. H. Mechanisms of failure of total knee arthroplasty. In W.N. Scott (Ed.), *The Knee*. St. Louis: Mosby, 1994. Pp. 1239-1249.

14. Murray D.W. and Rushton N. Macrophages stimulate bone resorption when they phagocytose particles. *J Bone and Joint Surg* **72-B**:988, 1990.

15. Glant T.T. and Jacobs J.J. Response of 3 murine macrophage populations to particulate debris – bone resorption in organ-cultures. *J Orthop Res* **12**:720, 1994.

16. Kapandji, I. A. The knee. In *The Physiology of the Joints*. London, N.Y.: Church Livingstone, 1970. Pp. 72-106.

17. Andriacchi, T. P., Stanwick, T. S., and Galante, J. O. Knee biomechanics and total knee replacement. *J Arthroplasty* **1**:211, 1986.

18. Schipplein, O. D. and Andriacchi, T. P. Interaction between active and passive knee stabilizers during level walking. *J Orthop Res* **9**:113, 1991.

19. Seireg, A. and Arvikar, R. J. The prediction of muscular load sharing and joint forces in the lower extremities during walking. *J Biomech* **8**:89, 1975.

20. Morrison, J. B. Bioengineering analysis of force actions transmitted by the knee joint. *Biomed Mater Eng* **3**:164, 1968.

21. Charnley, J. *Low friction arthroplasty of the hip*, New York: Springer, 1979.

22. Dumbleton, J. H. *Tribology of natural and artificial joints - Tribology Series 3*, Amsterdam: Elsevier, 1981.

23. Czichos, H. *Tribology - A systems approach to the science and technology of friction, lubrication and wear*, Amsterdam: Elsevier, 1978.

24. Fischer, A. Well-founded selection of materials for improved wear resistance. *Wear* **194**:238, 1996.

25. Habig, K. H. *Verschleiß und Härte von Werkstoffen*, München: Carl Hanser Verlag, 1980.

Chapter 2

1. Kapandji, I. A. The knee. In *The Physiology of the Joints*. London, N.Y.: Church Livingstone, 1970. Pp. 72-106.

2. Moore, K. L. *Clinically Oriented Anatomy*, 3rd ed., Baltimore: Williams & Wilkins, 1992.

3. Andriacchi, T. P. and Mikosz, R. P. Musculosceletal dynamics, locomotion and clinical applications. In V.C. Mow and W.C. Hayes (Eds.), *Basic Orthopaedic Biomechanics*. New York: Raven Press, 1991. Pp. 51-92.

4. Lafortune, M. A., Cavanagh, P. R., Sommer, H. J., and Kalenak, A. Three-dimensional kinematics of the human knee during walking. *J Biomech* **25**:347, 1992.

5. Weber, W. and Weber, E. *Mechanik der menschlichen Gehwerkzeuge*, Göttingen: 1836.

6. Strasser, H. *Lehrbuch der Muskel- und Gelenkmechanik*, Berlin: Springer Verlag, 1917.

7. Müller, W. *The Knee: Form, Function , and Ligament Reconstruction*, Berlin: Springer Verlag, 1983.

8. O'Connor, J. J., Shercliff, T., and Fitzpatrick, D. Geometry of the knee. In D.M Daniel, W.H. Akeson, and J.J. O'Connor (Eds.), *Knee Ligaments: Structure, Function, Injury, and Repair*. New York: Raven Press, 1990. Pp. 163-199.

9. O'Connor, J. J. and Zavatsky, A. Anterior cruciate ligament function in the normal knee. In D.W. Jackson (Ed.), *The Anterior Cruciate Ligament: Current and Future Concepts*. New York: Raven Press, Ltd., 1993. Pp. 39-52.

10. Draganich, L. F. *The influence of the cruciate ligaments, knee musculature and anatomy on knee joint loading*, University of Illinois at Chicago: Ph.D. Thesis, 1984.

11. Blankevoort, L. *Passive motion characteristics of the human knee joint*, University of Nijmegen: Ph.D. Thesis, 1991.

12. Draganich, L. F., Andriacchi, T. P., and Andersson, G. B. J. Interaction between intrinsic knee mechanics and the knee extensor mechanism. *J Orthop Res* **5**:539, 1987.

13. Andriacchi, T. P., Stanwick, T. S., and Galante, J. O. Knee biomechanics and total knee replacement. *J Arthroplasty* **1**:211, 1986.

14. Nisell, R. Mechanics of the Knee. *Acta Orthop Scand* **56** (Suppl. 216):1985.

15. Shaw, J. A. and Murray, D. G. The longitudinal axis of the knee and the role of the cruciate ligaments in controlling transverse rotation. *J. Bone Joint Surg* **56-A**:1603, 1974.

16. Rosenberg, A., Mikosz, R. P., and Mohler, C. G. Basic knee biomechanics. In W.N. Scott (Ed.), *The Knee*. St. Louis: Mosby, 1994. Pp. 75-94.

17. Gosh, P. and Taylor, T. K. F. The knee joint meniscus. A fibrocartilage of some distinction. *Clin Orthop* **224**:52, 1987.

18. Levy, M. The effect of lateral meniscectomy on motion of the knee. *J Bone Joint Surg* **71-A**:401, 1989.

19. Blankevoort, L., Huiskes, R., and de Lange, A. Recruitment of knee-joint ligaments. *J Biomech Eng* **113**:94, 1991.

20. Blankevoort, L., Huiskes, R., Kuiper, J. H., and Grootenboer, H. J. Articular contact in a three-dimesional model of the knee. *J Biomech* **24**:1019, 1991.

21. Mommersteeg, D. *Human Knee Ligaments*, University of Nijmegen: Ph.D. Thesis, 1994.

22. Moilanen, T. and Freeman, M. A. R. Point - counterpoint of total knee arthroplasty: the case for resection of the posterior cruciate ligament. *J Arthroplasty* **10**:564, 1995.

23. Dorr, L. D. and et al., Functional comparison of posterior crucuiate-retained versus cruciate-sacrificed total knee arthroplasty. *Clin Orthop* **236**:36, 1988.

24. Andriacchi, T. P., Galante, J. O., and Fermier, R. W. The influence of total knee replacement design on function during walking and stair climbing. *J Bone Joint Surg* **64**:1328, 1982.

25. Matthews, L. S., Sonstegard, D. A., and Henke, J. A. Load bearing characteristics of the patellofemoral joint. *Acta Orthop Scand* **48**:511, 1977.

26. Van Eijden, T. M., De Boer, W., and Weijs, W. A. The orientation of the distal part of the quadriceps femoris muscle as a function of the knee flexion-extension angle. *J Biomech* **18**:803, 1985.

27. Ahmed, A. M., Burke, D. L., and Hyder, A. Force analysis of the patellar mechanism. *J Orthop Res* **5**:69, 1987.

28. Andriacchi, T. P., Galante, J. O., and Draganich, L. F. Relationship between knee extensor mechanics and function following total knee replacement. In L.D. Dorr (Ed.), *The Knee*. Baltimore: University Park Press, 1985. Pp. 83-94.

29. Li, E. and Ritter, M. A. Point - counterpoint of total knee arthroplasty: The case for retention of the posterior cruciate ligament. *J Arthroplasty* **10**:560, 1995.

30. Andriacchi, T. P. and Strickland, A. B. Gait analysis as a tool to assess joint kinetics. In N. Berme, A.E. Engin, and K.M. Correia da Silva (Eds.), *Biomechanics of Normal and Pathological Human Articulating Joints*. NATO ASI Series, No. 93, 1985. Pp. 83-102.

31. Olney, S. J. and Winter, D. A. Predictions of knee and ankle moments of force in walking from EMG and kinematic data. *J Biomech* **18**:9, 1985.

32. Sepulveda, F., Wells, D. M., and Vaughan, C. L. A neural network presentation of electromyography and joint dynamics in human gait. *J Biomech* **26**:101, 1993.

33. Cholewicki, J. and McGill, S. M. EMG assisted optimization: a hybrid approach for estimating muscle forces in an indeterminate biomechanical model. *J Biomech* **27**:1287, 1994.

34. Perry, J. The mechanics of gait. In V. Wright and E.L. Radin (Eds.), *Mechanics of human joints: physiology, pathophysiology, and treatment*. New York: Marcel Dekker, Inc., 1998. Pp. 83-107.

35. Berchuck, M., Andriacchi, T. P., Bach, B. R., and Reider, B. Gait adaptions by patients who have a deficient anterior cruciate ligament. *J Bone Joint Surg* **72-A**:871, 1990.

36. Andriacchi, T. P. Dynamics of pathological motion: applied to the anterior cruciate deficient knee. *J Biomech* **23**:99, 1990.

37. Andriacchi, T. P. Functional evaluation of normal and ACL-deficient knee using gait analysis technique. In D.W. Jackson (Ed.), *The anterior cruciate ligament: current and future concepts.* New York: Raven Press, 1993. Pp. 153-159.

38. Schipplein, O. D. and Andriacchi, T. P. Interaction between active and passive knee stabilizers during level walking. *J Orthop Res* **9**:113, 1991.

39. Andriacchi, T. P. Gait analysis and total knee replacement. In J.N. Insall, W.N. Scott, and G.R. Scuderi (Eds.), *Current Concepts in Primary and Revision Total Knee Arthroplasty.* Philadelphia: Lippincott-Raven Publishers, 1996. Pp. 29-36.

40. Hodge, W. A., Carlson, K. L., Fijan, S. M., Burgess, R. G., Riley, P. O., Harris, W. H., and Mann, R. W. Contact pressures from an instrumented hip endoprosthesis. *J Bone Joint Surg* **71-A**:1378, 1889.

41. Bergmann, G., Graichen, F., and Rohlmann, A. Hip joint loading during walking and running measured in two patients. *J Biomech* **26**:969, 1993.

42. Davy, D. T., Kotzar, G. M., Brown, R. H., Heiple, K. G., Goldberg, V. M., Berilla, J., and Bursteing, A. H. Telemetric force measurements across the hip after total hip arthroplasty. *J Bone Joint Surg* **70-A**:45, 1988.

43. Rydell, N. W. *Forces acting on the femoral head prosthesis: A study of strain gauge supplied prostheses in living persons*, Goteborg: 1966.

44. Taylor, S. J., Walker, P. S., Perry, J., Cannon, S. R., and Woledge, R. The forces in the distal femur and knee during different activities measured by telemetry. *Trans Orthop Res Soc* **22**:259, 1997.(Abstract)

45. Wismans, J., Veldpaus, F., and Janssen, J. A three-dimensional mathematical model of the knee-joint. *J Biomech* **13**:677, 1980.

46. Andriacchi, T. P., Mikosz, R. P., and Hampton, S. J. Model studies of the stiffness characteristics of the human knee joint. *J Biomech* **16**:23, 1983.

47. Morrison, J. B. Bioengineering analysis of force actions transmitted by the knee joint. *Biomed Mater Eng* **3**:164, 1968.

48. Morrison, J. B. Function of the knee joint in various activities. *Biomed Mater Eng* **4**:573, 1969.

49. Seireg, A. and Arvikar, R. J. The prediction of muscular load sharing and joint forces in the lower extremities during walking. *J Biomech* **8**:89, 1975.

50. Crowninshield, R. D. and Brand, R. A. A physiologically based criteria of muscle force prediction in locomotion. *J Biomech* **14**:793, 1981.

51. Patriarco, A. G., Mann, R. W., Simon, S. R., and Mansour, J. M. An evaluation of the approaches of optimization models in the prediction of muscle forces during human gait. *J Biomech* **14**:513, 1981.

52. Hill, A.V. The mechanics of the active muscle. *Proc Roy Soc Lond* **38**:57, 1953.

53. Mow,V.C., Fithian, D.C., and Kelly, M.A. Fundamentals of articular cartilage and meniscus biomechanics In J.W. Ewing (Ed.) *Articular Ccartilage and Knee Joint Function: Basic Science and Arthroscopy*, Raven Press, New York, 1990. Pp. 1-18.

54. Chao, E. Y. S. Justification of triaxial goniometer for the measurement of joint rotation. *J. Biomech* **13**:989, 1980.

55. *DTV Atlas der Anatomie; Band 1*, Stuttgart: Georg Thieme Verlag, 1986.

56. Rosenberg, A., Mohler, C. G., and Mikosz, R. P. Gait analysis and its relationship to knee function. In W.N. Scott (Ed.), *The Knee*. St. Louis: Mosby, 1994. Pp. 95-105.

57. O'Connor, J. J. and Zavatsky, A. Kinematics and mechanics of the cruciate ligaments of the knee. In V.C. Mow, A. Ratelitte, and S.L-Y. Woo (Eds.), *Biomechanics of Diathrodial Joints, Vol.II*. New York: Springer, 1990. Pp. 197-241.

58. Morrison, J. B. The mechanics of the knee joint in relation to normal walking. *J Biomech* **3**:51, 1970.

Chapter 3

1. Dowson, D. *History of Tribology*, London: Longman, 1975.

2. Jost, P. H. Tribology - Origin and future. *Wear* **136**:1, 1990.

3. Zum Gahr, K. -H. *Microstructure and Wear of Materials*, Amsterdam: Elsevier, Tribology Series Vol.10, 1987.

4. DIN 50 320, Verschleiß - Begriffe, Systemanalyse von Verschleißvorgängen, Gliederung des Verschleißgebietes. *Deutsche Normen* 1979.

5. Godet, M. The third-body approach: a mechanical view of wear. *Wear* **100**:437, 1984.

6. Czichos, H. *Tribology - A systems approach to the science and technology of friction, lubrication and wear*, Amsterdam: Elsevier, 1978.

7. Habig, K. H. *Verschleiß und Härte von Werkstoffen*, München: Carl Hanser Verlag, 1980.

8. Uetz, H., Sommer, K., and Khosrawi, M. A. Übertragbarkeit von Versuchs- und Prüfergebnissen bei abrasiver Verschleißbeanspruchung auf Bauteile. *VDI-Berichte* **354**:107, 1979.

9. Uetz, H. and Wiedemeyer, J. *Tribologie der Polymere*, München: Carl Hanser Verlag, 1984.

10. Fischer, A. *Einfluß der Temperatur auf das tribologische Verhalten metallischer Werkstoffe*, Düsseldorf: VDI Verlag, 1994.

11. Petermann, J. and Broza, G. Gefüge von Polymeren. *Verbindungstechnik in der Elektronik* **1**:14, 1990.

12. Black, J. *Orthopaedic Biomaterials in Research and Practice*, New York: Chirchill Livingstone, 1988.

13. Rabinowicz, E. *Friction and wear of materials*, New York: John Wiley & Sons, 1995.

14. Kagiya, V. T., Takemoto, K., and Hagiwara, M. Elementary reactions in polymer degradation. *Journal of Applied Polymer Science* **35**:95, 1979.

15. Jansson, J. F. and Terselius, B. Mechano-chemical phenomena in polymers. *Journal of Applied Polymer Science* **35**:455, 1979.

16. Dagnall, H. *Exploring Surface Texture*, Leicester: Rank Taylor Hobson, 1986.

17. Hutchings, I. M. *Tribology: friction and wear of engineering materials*, London: Edward Arnold, 1992.

18. Greenwood, J. A. and Williamson, J. B. P. Contact of nominally flat rough surfaces. *Proc Roy Soc Lond* **A295**:300, 1966.

19. Yamaguchi, Y. *Tribology of plastic materials (Tribology Series 16)*, Amsterdam: Elsevier, 1990.

20. Bowden, F. P. and Tabor, D. *The Friction and Lubrication of Solids*, Oxford: Clarendon Press, 1964.

21. Reynolds, O. On rolling friction. *Phil Trans Roy Soc Lond* **166**: 155, 1876.

22. Kragelsky, I. V., Dobychin, M. N., and Kombalov, V. S. *Friction and Wear: Calculation Methods*, Oxford: Pergamon Press, 1982.

23. Johnsen, K. L. *Contact Mechanics*, Cambridge: Cambridge University Press, 1985.

24. Dowson, D., El-Hady Diab, M. M., Gillis, B. J., and Atkinson, J. R. Influence of counterface topography on the wear of ultra high molecular weight polyethylene under wet or dry conditions. In Lieng-Huang Lee (Ed.), *Polymer Wear and its Control*. Washington, DC.: American Chemical Society, 1985. Pp. 171-187.

25. O'Kelly, J., Unsworth, A., Dowson, D., and Wright, V. An experimental study of friction and lubrication in hip prostheses. *Eng Med* **8**:153, 1979.

26. Dumbleton, J. H. *Tribology of natural and artificial joints - Tribology Series 3*, Amsterdam: Elsevier, 1981.

27. Waldman, S. D. and Bryant, J. T. Compressive stress relaxation behaviour of irradiated ultra high molecular weight polyethylene at 37°C. *J Appl Biomater* **5**:333, 1994.

28. Charnley, J. *Low friction arthroplasty of the hip*, New York: Springer Verlag, 1979.

29. Birnkraut, H.W. Synthesis of UHMW-PE. In H.-G. Willert, G.H. Buchhorn and P. Eyerer *Ultra-High Molecular Weight Polyethylene as Biomaterial in Orthopedic Surgery*. Toronto, Hogrefe & Huber Publishers, 1985. Pp.3-5.

30. Eyerer, P., Kurth, M., McKellop, H.A. and Mittlmeier, T. Characterization of UHMWPE hip cups run on joint simulators. *J Biomed Mat Res* **21**:275, 1987.

31. Encyclopaedia of Polymer Science and Engineering, New York: Wiley-Interscience, 1985. Vol.6., p.490.

32. Murakami, T. The lubrication in natural synovial joints and joint prostheses. *JSME International Journal* **33**:465, 1990.

33. Davidson, J. A., Poggie, R. A., and Mishra, A. K. Abrasive wear of ceramic, metal, and UHMWPE bearing surfaces from third-body bone, PMMA bone cement, and titanium debris. Biomed Mater Eng 4:213, 1994.

34. Premnath, V., Harris, W.H., Jasty, M. And Merril, E.W. Gamma sterilization of UHMWPE articular implants: an analysis of the oxidation problem. *Biomaterials* **17**: 1741, 1996.

35. Tabor, D. Lubrication and Wear. In Matijevic E. (Ed.), *Surface and Colloid Science* (Vol. 5) New York: John Wiley, 1977. Pp. 245-312.

Chapter 4

1. Knutson, K., Lewold, S., Robertsson, O., and Lidgren, L. The swedish knee arthroplasty register: A nation wide study of 30,003 knees 1976-1992. *Acta Orthop Scand* **65**:375, 1994.

2. Dorr, L. D. and Serocki, J. H. Mechanisms of failure of total knee arthroplasty. In W.N. Scott (Ed.), *The Knee*. St. Louis: Mosby, 1994. Pp. 1239-1249.

3. Harris, W. H. The problem is osteolysis. *Clin Orthop* **311**:46, 1995.

4. Goldring, S. R., Clark, C. R., and Wright, T. M. Editorial. The problem in total joint arthroplasty: Aseptic loosening. *J Bone Joint Surg* **75-A**:799, 1993.

5. Goodman, S. B., Fornasier, V. L., and Kei, J. The effect of bulk versus particulate ultra-high-molecular-weight polyethylene on bone. *J Arthroplasty* **Suppl**:41, 1988.

6. Dowson, D. Editorial. In D. Dowson (Ed.), *Advances in Medical Tribology*. Bury St. Edmunds: Mechanical Engineering Publications Ltd., 1998.

7. Peters, P. C., Engh, G. A., Dwyer, K. A., and Vinh, T. N. Osteolysis after total knee arthroplasty without cement. *J Bone Joint Surg* **74-A**:864, 1992.

8. Robinson, E. J., Mulliken, B. D., Bourne, R. B., Rorabeck, C. H., and Alvarez, C. Catastrophic osteolysis in total knee replacement. *Clin Orthop* **321**:98, 1995.

9. Willert, H. G., Bertram, H., and Bushhorn, G. H. Osteolysis in alloarthroplasty of the hip. The role of ultra-high molecular weight polyethylene wear particles. *Clin Orthop* **258**:95, 1990.

10. Jones, L. C. and Hungerford, D. S. Cement desease. *Clin Orthop* **225**:192, 1987.

11. Maloney, W. J., Jasty, M., Harris, W. H., Galante, J. O., and Callaghan, J. J. Endosteal erosion in association with stable uncemented femoral components . *J Bone Joint Surg* **72-A**:1025, 1990.

12. Amstutz, H. C., Campbell, P., Kossovsky, N., and Clarke, I. C. Mechanism and clinical significance of wear debris-induced osteolysis. *Clin Orthop* **276**:7, 1992.

13. Willert, H. G. and Buchhorn, G. H. Particle disease due to wear of ultrahigh molecular weight polyethylene. In B.F. Morrey (Ed.), *Biological, Material, and Mechanical Considerations of Joint Replacement*. New York: Raven Press, Ltd., 1993. Pp. 87-102.

14. Friedman, R. J., Black, J., Galante, J. O., Jacobs, J. J., and Skinner, H. B. Current concepts in orthopaedic biomaterials and implant fixation. *J Bone Joint Surg* **75-A**:1086, 1993.

15. Shanbhag, A. S., Bailey, H. O., Woo, S. L-Y, and Rubash, H. E. Characterization and comparison of UHMWPE wear debris retrieved from total hip and knee prosthesis. *Trans Orthop Res Soc* **42**:467, 1996.

16. Horikoshi, M., Macaulay, W., Booth, R. E., Crossett, L. S., and Rubash, H. E. Comparison of interface membranes obtained from failed cemented and cementless hip and knee prostheses. *Clin Orthop* **309**:69, 1994.

17. Bosco, J., Benjamin, J., and Wallace, D. Quantitative and qualitative analysis of polyethylene wear particles in synovial fluid of patients with total knee arthroplasty. *Clin Orthop* **309**:11, 1994.

18. Schmalzried, T. P., Campbell, P., Brown, I. C., Schmitt, A. K., and Amstutz, H. C. Polyethylene wear particles generated in vivo by total knee replacements compared to total hip replacements. *Trans Orthop Res Soc* **20**:163, 1995.

19. Hirakawa, K., Bauer, T. W., Stulberg, B. N., and Wilde, A. H. Comparison and quantification of wear debris of failed total hip and total knee arthroplasty. *J Biomed Mater Res* **31**:257, 1996.

20. Murray, D. W. and Rushton, N. Macrophages stimulate bone resorption when they phagocytose particles. *J Bone Joint Surg* **72-B**:988, 1990.

21. Shanbhag, A. S., Jacobs, J. J., Black, J., Galante, J. O., and Glant, T. T. Macrophage/ particle interactions: Effect of size, composition and surface area. *J Biomed Mater Res* **28**:81, 1994.

22. Shanbhag, A. S., Jacobs, J. J., Black, J., Galante, J. O., and Glant, T. T. Human monocyte response to particulate biomaterials generated in vivo and in vitro. *J Orthop Res* **13**:792, 1995.

23. Jiranek, W. A., Machado, M., Jasty, M., Jevsevar, D., Wolfe, H. J., Goldring, S. R., Goldberg, M. J., and Harris, W. H. Production of cytokines around loosened cemented acetabular components. *J Bone Joint Surg* **75-A**:863, 1993.

24. Takamura, K., Hayashi, K., Ishinishi, K., Yamada, T., and Sugioka, Y. Evaluation of carcinogenicity and chronic toxicity associated with orthopaedic implant in mice. *J Biomed Mater Res* **28**:583, 1994.

25. Willert, H. G. and Semlitsch, M. Tissue reactions to plastic and metallic wear products of joint endoprostheses. *Clin Orthop* **333**:4, 1996.

26. Wilson, C. M. G. *In vitro biocompatability evaluation and morphological description of fretting wear debris from orthopaedic implant materials*. Aberystwyth, Ph.D.-Thesis, University of Wales, 1999.

27. Furman, B. D., Ritter, M. A., Perone, J. B., Furman, G. L., and Li, S. Effect of resin type and manufacturing method on UHMWPE oxidation and quality at long aging and implant times. *Trans Orthop Res Soc* **22**:92, 1997.

28. Mayor, M. B., Wrona, M., Collier, J. P., and Jensen, R. E. The role of polyethylene quality in the failure of tibial knee components. *Trans Orthop Res Soc* **39**:292, 1993.

29. Wrona, M., Mayor, M. B., Collier, J. D., and Jensen, R. D. The correlation between fusion defects and damage in tibial polyethylene bearings. *Clin Orthop* **299**:92, 1994.

30. Weightman, B. and Light, D. A comparison of RCH 1000 and Hi-Fax 1900 ultra-high molecular weight polyethylenes. *Biomaterials* **6**:177, 1985.

31. Walker, P. S., Blunn, G. W., and Lilley, P. A. Wear testing of materials and surfaces for total knee replacement. *J Biomed Mater Res* **33**:159, 1996.

32. Eyerer, P. Kunststoffe in der Gelenkendoprothetik. *Zeitschrift für Werkstofftechnik* **17**:384, 1986.

33. Rose, R. M., Goldfarb, H. V., Ellis, E., and Crugnola, A. M. On the pressure dependence of the wear of ultrahigh molecular weight polyethylene. *Wear* **92**:99, 1983.

34. Li, S. and Burstein, A. H. Current concepts review: ultra-high molecular weight polyethylene. *J Bone Joint Surg* **76-A**:1080, 1994.

35. Barnetson, A. and Hornsby, P. R. Observations on the sintering of ultra-high molecular weight polyethylene (UHMWPE) powders. *Journal of Materials Science* **14**:80, 1995.

36. Pienkowski, D., Jacob, R., Hoglin, D., Saum, K., Kaufer, H., and Nicholls, P. J. Low-voltage scanning electron microscopic imaging of ultrahigh-molecular-weight polyethylene. *J Biomed Mater Res* **29**:1167, 1995.

37. Lewis, G. Polyethylene wear in total hip and knee arthroplasties. *J Biomed Mater Res (Applied Biomaterials)* **38**:55, 1997.

38. Lykins, M. D. and Evans, M. A. A comparison of extruded and molded UHMWPE. *Trans Soc Biomat* **21**:385, 1995.

39. Huber, J., Plitz, W., Walter, A., and Refior, H. J. Vergleichende tribologische Untersuchungen von Chirulen®, Hylamer®, und Enduron® gepaart mit Al2O3. *Orthopäde* **26**:125, 1997.

40. Huber, J., Walter, A., Plitz, W., and Refior, H. J. The influence of the manufacturing process on creep and wear properties of UHMWPE. *Biomedizinische Technik* **40**:88, 1995.

41. Bankston, A. B., Keating, E. M., Ranawat, C., Faris, P. M., and Ritter, M. A. Comparison of polyethylene wear in machined versus molded polyethylene. *Clin Orthop* **317**:37, 1995.

42. Tanner, M. G., Whiteside, L. A., and White, S. E. Effect of polyethylene quality on wear in total knee arthroplasty. *Clin Orthop* **317**:83, 1995.

43. Jasty, M., James, S. P., Bragdon, C. R., Elder, J., Lowenstein, J., and Harris, W. H. Microstructural ans ultrastructural features which may influence the wear of ultra high molecular weight polyethylenes. *Trans Orthop Res Soc* **19**:586, 1994.

44. Sun, D. C., Stark, C., and Dumbleton, J. H. Charakterization and comparison of compression molded and machined UHMWPE components. *Trans Orthop Res Soc* **19**:173, 1994.

45. Jerosch, J., Fuchs, S., Schmidt, Th., and Reichelt, R. Scanning electron microscopic Studies of morphological changes in UHMWPE induced by various forms of machining. *Biomedizinische Technik* **41**:138, 1996.

46. Bristol, R. E., Fitzpatrick, D. C., Brown, T. D., and Callaghan, J. J. Non-uniformity of contact stress on polyethylene insert in total knee arthroplasty. *Clinical Biomechanics* **11**:75, 1996.

47. Wright, T. M., Rimnac, C. M., Stulberg, S. D., Mintz, L., Tsao, A. K., Klein, R. W., and McCrae, C. Wear of polyethylene in total joint replacements: Observations from retrieved PCA knee implants. *Clin Orthop* **276**:126, 1992.

48. Bloebaum, R. D., Nelson, K., Dorr, L. D., Hofmann, A. A., and Lyman, D. J. Investigation of early surface delamination observed in retrieved heat-pressed tibial inserts. *Clin Orthop* **269**:120, 1991.

49. Nusbaum, H. J. and Rose, R. M. The effects of radiation sterilization on the properties of ultrahigh molecular weight polyethylene. *J Biomed Mater Res* **13**:557, 1979.

50. Jones, W. R. and Hady, F. H. Effect of sterilization irradiation on friction and wear of ultrahigh-molecular-weight polyethylene. *NASA Technical Publications* **1462**:4, 1979.

51. Buchalla, R., Schuettler, C., and Boegl, K. W. Radiation sterilization of medical devices. Effects of ionizing radiation on ultra-high molecular-weight polyethlene. *Radiat Phys Chem* **46**:579, 1995.

52. Bostrom, M. P., Bennett, A. P., Rimnac, C. M., and Wright, T. M. The natural history of ultra high molecular weight polyethylene. *Clin Orthop* **309**:20, 1994.

53. Otfinowski, J. and Pawelec, A. Changing crystallinity of polyethlene in the acetabular cups of weller hip prostheses. *J Bone Joint Surg* **77-B**:802, 1995.

54. Sutula, L. C., Collier, J. P., Saum, K. A., Currier, B. H., Currier, J. H., Sanford, M. W., Mayor, M. B., Wooding, R. E., Sperling, D. K., Williams, I. R., Kasprzak, D. J., and Surprenant, V. A. Impact of gamma sterilization on clinical performance of polyethlene in the hip. *Clin Orthop* **319**:28, 1995.

55. Premnath, V., Harris, W. H., Jasty, M., and Merrill, E. W. Gamma sterilization of UHMWPE articular implants: an analysis of the oxidation problem. *Biomaterials* **17**:1741, 1996.

56. Birkinshaw, C., Buggy, M., and Daly, S. Mechanisms of aging in irradiated polymers. *Polymer Degrad. and Stabil.* **2**:285, 1988.

57. Muratoglu, O. K., Bragdon, C. R., Jasty, M., and Harris, W. H. An oxidation mechanism in gamma sterilized ultra-high-molecular-weight-polyethylene components. *Proc. Europ. Soc. Biomech.* **10**:184, 1996.

58. Jahan, M. S., Wang, C., Schwartz, G., and Davidson, J. A. Combined chemical and mechanical effects on free radicals in UHMWPE joints during implantation. *J Biomed Mater Res* **25**:1005, 1991.

59. Rimnac, C. M., Burstein, A. H., Carr, J. M., Klein, R. W., Wright, T. M., and Betts, F. Chemical and mechanical degradatin of UHMWPE: Report of the Development of an in vitro test. *J. Appl. Biomater.* **5**:17, 1994.

60. Grood, E. S., Shastri, S., and Hopsen, C. N. Analysis of retrieved implants: Crystallinity changes in ultrahigh molecular weight polyethylene. *J Biomed Mater Res* **16**:399, 1982.

61. Fisher, J., Hailey, J. L., Chan, K. L., Shaw, D., and Stone, M. The effect of ageing following irradiation on the wear of UHMWPE. *Trans Orthop Res Soc* **20**:120, 1995.

62. Furman, B. D., Lelas, J., McNulty, D., Smith, T., and Li, S. Kinetics, chemistry and calibration of UHMWPE accelerated aging methods. *Trans Orthop Res Soc* **23**:102, 1998.

63. Bell, C. J., Simons, J., King, P., Walker, P. S., and Blunn, G. W. Is oxidation of ultra high molecular weight polyethylene the main cause of delamination wear in total knee replacement? *Trans Orthop Res Soc* **22**:96, 1997.

64. Streicher, R. M. Sterilization and long-term aging of medical-grade UHMWPE. *Radiat Phys Chem* **46**:893, 1995.

65. Jerosch, J., Fuchs, S., Liljenqvist, U., and Haftka, A. Influence of different sterilization techniques on the oxidation of UHMWPE. *Biomedizinische Technik* **40**:296, 1995.

66. Eyerer, P. and Ke, Y. C. Property changes of UHMW polyethylene hip cup endoprostheses during implantation. *J Biomed Mater Res* **18**:1137, 1984.

67. Goodman, S. and Lidgren, L. Polyethylene wear in knee arthroplasty: A review. *Acta Orthop Scand* **63** (3):358, 1992.

68. Peterson, C. D., Hillberry, B. M., and Heck, D. A. Component wear of total knee prostheses using Ti-6Al-4V, titanium nitride coated Ti-6Al-4V, and cobalt-chromium-molybdenum femoral components. *J Biomed Mater Res* **22**:887, 1988.

69. McKellop, H., Clarke, I. C., Markolf, K. L., and Amstutz, H. C. Friction and wear properties of polymer, metal, and ceramic prosthetic joint materials evaluated on a multichannel screening device. *J Biomed Mater Res* **15**:619, 1981.

70. Kraay, M. J., Goldberg, V. M., Brown, S. A., and Merritt, K. Retrieval analysis of Ti 6Al-4V Miller-Gallante total knee replacements. In S.A. Brown and J.E. Lemons (Eds.), *Medical applicatons of titanium and its alloys: The material and biological issues, ASTM STP 1272*. American Society for Testing and Materials, 1996. Pp. 409-416.

71. Davidson, J. A., Mishra, A. K., Poggie, R. A., and Wert, J. J. Sliding friction and UHMWPE wear comparison between cobalt alloy and zirconia surfaces. *Trans Orthop Res Soc* **38**:404, 1992.

72. White, S. E., Whiteside, L. A., McCarthy, D. S., Anthony, M., and Poggie, R. A. Simulated knee wear with cobalt chromium and oxidized zirconium knee femoral components. *Clin Orthop* **309**:176, 1994.

73. Davidson, J.A., Mishra, A.K., and Poggie, R.A., The effect of bone, bone cement, and metal debris on the change in roughness of ceramic and metal orthopaedic implant bearing surfaces. *Eurotrib '93*, 6th International Congress of Tribology, Budapest, Hungary 1993. Pp. 157-164.

74. Weightman, B. and Light, D. The effect of surface finish of alumina and stainless steel on the wear rate of UHMW polyethylene. *Biomaterials* **7**:20, 1986.

75. Lloyd, A. I. and Noël, R. E. The effect of counterface surface roughness on the wear of UHMWPE in water and oil-in-water emulsion. *Tribology International* **301**:83, 1988.

76. McNie, C. M., Barton, D. C., Ingham, E., Stone, M. H., Tipper, J. L., and Fisher, J. Experimental and theoretical predictions of wear particle generation in UHMWPE due to microscopic femoral counterface asperities. *Trans Orthop Res Soc* **23**:74, 1998.

77. Wang, A., Polineni, V. K., Stark, C., and Dumbleton, J. H. True effect of femoral head surface roughness on the wear of UHMWPE acetabular cups. *Trans Orthop Res Soc* **23**:76, 1998.

78. Niemann, J. C., Brown, T. D., Pedersen, D. R., and Callaghan, J. J. The effects of localized regions of head roughening on variability of UHMWPE wear in THA. *Trans Orthop Res Soc* **23**:77, 1998.

79. Dowson, D., El-Hady Diab, M. M., Gillis, B. J., and Atkinson, J. R. Influence of counterface topography on the wear of ultra high molecular weight polyethylene under wet or dry conditions. In L.-H. Lee (Ed.), *Polymer Wear and its Control*. Washington, D.C.: American Chemical Society, 1985. Pp. 171-187.

80. Fisher, J., Dowson, D., Hamdzah, H., and Lee, H. L. The effect of sliding velocity on the friction and wear of UHMWPE for use in total artificial joints. *Wear* **175**:219, 1994.

81. Marcus, K., Ball, A., and Allen, C. The effect of grinding direction on the nature of the transfer film formed during the sliding wear of UHMPE against stainless steel. *Wear of Materials* **ASME**:571, 1991.

82. Dowson, D., Taheri, S., and Wallbridge, N. C. The role of counterface imperfections in the wear of polyethylene. *Wear of Materials* **ASME**:415, 1987.

83. Hailey, J. L., Ingham, E., Fisher, J., Dowson, D., and Wroblewski, B. M. Influence of counterface roughness on the morphology of polyethylene wear debris. *Trans Orthop Res Soc* **20**:763, 1995.

84. McKellop, H., Clarke, I. C., Markolf, K. L., and Amstutz, H. C. Wear characteristics of UHMW polyethylene: A method for accurately measuring extremely low wear rates. *J Biomed Mater Res* **12**:895, 1978.

85. McKellop, H. Wear of artificial joint materials II. Twelve-channel wear-screening device: correlation of experimental and clinical results. *Eng Med* **10**:123, 1981.

86. Liao, X., Benya, P. D., Lu, B., and McKellop, H. Stability of serum as a lubricant in wear simulator tests of prosthetic joints. *Proc. 5th World Biomat. Congress* **5**:871, 1996.

87. Hall, R. M. and Unsworth, A. Review: friction in hip prostheses. *Biomaterials* **18**:1017, 1997.

88. Campbell, P. A., Liu, T. S., Schmalzried, T. P., McKellop, H., Gonsalves, R., and Amstutz, H. C. Characteristics of fluids around total hip and total knee replacements. *Trans Soc Biomat* **23**:130, 1997.

89. McKellop, H. A., Liu, T-S., Campbell, P. A., Schmalzried, T. P., and Amstutz, H. C. Lubricating properties of joint fluids around total hip and knee replacements. *Trans Orthop Res Soc* **23**:74, 1998.

90. Rostocker, W. and Galante, J. O. Contact pressure dependence of wear rates of ultra high molecular weight polyethylene. *J Biomed Mater Res* **13**:957, 1979.

91. Walker, P. S., Ben-Dov, M., Askew, M. J., and Pugh, J. The deformation and wear of plastic components in artificial knee joints - an experimental study. *Eng Med* **10**:33, 1981.

92. Treharne, R. W., Young, R. W., and Young, S. R. Wear of artificial joint materials III: Simulation of the knee joint using a computer-controlled system. *Eng Med* **10** (3):137, 1981.

93. Fusaro, R. L. Effect of sliding speed and contact stress on tribological properties of ultra-high-molecular-weight polyethylene. *NASA Technical Publications* **2059**:1, 1982.

94. Pruitt, L., Koo, J., Rimnac, C. M., Suresh, S., and Wright, T. M. Cyclcic compressive loading results in fatigue cracks in ultra high molecular weight polyethylene. *J Orthop Res* **13**:143, 1995.

95. Maxian, T. A., Brown, T. D., Pedersen, D. R., and Callaghan, J. J. A sliding-distance-coupled finite element formulation for polyethylene wear in total hip arthroplasty. *J Biomech* **29**:687, 1996.

96. Briscoe, B. J. and Stolarski, T. A. The influence of contact zone kinematics on wear process of polymers. *Wear* **149**:233, 1991.

97. Ramamurti, B. S., Bragdon, C. R., O'Conner, D. O., Lowenstein, J. D., Jasty, M., Estock, D. M., and Harris, W. H. Loci of movement of selected points on the femoral head during normal gait. *J Arthroplasty* **11**:845, 1996.

98. Blunn, G. W., Walker, P. S., Joshi, A., and Hardinge, K. The dominance of cyclic sliding in producing wear in total knee replacements. *Clin Orthop* **273**:253, 1991.

99. Bragdon, C. R., O'Conner, D. O., Lowenstein, J. D., Jasty, M., and Syniuta, W. D. The importance of multidirectional motion on the wear of polyethylene. *Proc Inst Mech Eng [H]* **210**:157, 1996.

100. Wimmer, M. A., Nassutt, R., Lampe, F., Schneider, E., and Morlock, M. M. A new screening method designed for wear analysis of bearing surfaces used in total hip arthroplasty. In J. Jacobs, T. Cendrowska, and P. Speiser (Eds.), *Alternative Bearing Surfaces in Total Joint Replacement, ASTM STP 1346*. American Society for Testing and Materials, 1998.

101. Bragdon, C. R., O'Conner, D. O., Lowenstein, J. D., Jasty, M., and Harris, W. H. Development of a new pin on disc testing machine for evaluating polyethylene wear. *Proc Euro Soc Biomech* **10**:114, 1996.

102. Hood, R. W., Wright, T. M., and Burstein, A. H. Retrieval analysis of total knee prostheses: A method and its application to 48 total condylar prostheses. *J Biomed Mater Res* **17**:829, 1983.

103. Jensen, R. E., Collier, J. P., Mayor, M. B., and Surprenant, V. A. The role of polyethylene uniformity and patient characteristics in the wear of tibial knee components. *Trans Orthop Res Soc* **38**:328, 1992.

104. Engh, G. A., Dwyer, K. A., and Hanes, C. K. Polyethylene wear of metal-backed tibial components in total and unicompartmental knee prostheses. *J Bone Joint Surg (Br)* **74-B**:9, 1992.

105. Lavernia, C. J., Guzman, J. F., Kabo, M., Krackow, K., and Hungerford, D. Polyethylene wear in autopsy retrieved fully functional totol knee replacement. *Trans Soc Biomat* **20**:82, 1994.

106. Schmalzried, T.P., Szuszczewicz, E.S., Northfield, M.R., Akizuki, K.H., Frankel, R.E., Belcher, G., and Amstutz, H.C. Quantitative assessment of activity in joint replacement patients. *Trans Orthop Res Soc* **22**:284, 1997.

107. Seedhom, B. B. and Wallbridge, N. C. Walking activities and wear of prostheses. *Annals of the Rheumatic Diseases* **44**:838, 1985.

108. Andriacchi, T. P., Galante, J. O., and Fermier, R. W. The influence of total knee replacement design on function during walking and stair climbing. *J Bone Joint Surg* **64**:1328, 1982.

109. Schipplein, O. D. and Andriacchi, T. P. Interaction between active and passive knee stabilizers during level walking. *J Orthop Res* **9**:113, 1991.

110. Hilding, M. B., Lanshammer, H., and Ryd, L. Knee joint loading and tibial component loosening. *J Bone Joint Surg* **78-B**:66, 1996.

111. Banks, S. A., Markovich, G. D., and Hodge, W. A. Total knee replacement mechanics during gait. *Trans Orthop Res Soc* **22**:263, 1997.

112. Tarnowski, L. E., Berger, R. A., Andriacchi, T. P., Toney, M. K., Galante, J. O., and Rosenberg, A. G. Patterns of A-P Movement of the femur during stairclimbing in cruciate retaining and substituting totol knee replacement. *Proc Am Soc Biomech* **22**, 1998.

113. Andriacchi, T. P. and Galante, J. O. Retention of the posterior cruciate ligament in total knee arthroplasty. *J Arthroplasty* **Suppl**:13, 1988.

114. Bartel, D. L., Burstein, A. H., Toda, M. D., and Edwards, D. L. The effect of conformity and plastic thickness on contact stresses in metal-backed plastic implants. *Journal of Biomechanical Engineering* **107**:193, 1985.

115. Bartel, D. L., Rawlinson, J. J., Burstein, A. H., Ranawat, C. S., and Flynn, W. F. Stresses in polyethlene components of contemporary total knee replacements. *Clin Orthop* **317**:76, 1995.

116. Soudry, M., Walker, P. S., Reilly, D. T., Kurosawa, H., and Sledge, C. B. Effects of total knee replacement design on femoral-tibial contact conditions. *J Arthroplasty* **1**:35, 1986.

117. Bartel, D. L., Bicknell, V. L., and Wright, T. M. The effect of conformity, thickness and material on stresses in ultra-high molecular weight components for total joint replacement. *J Bone Joint Surg* **68-A**:1041, 1986.

118. Andriacchi, T. P. Dynamics of knee malalignment. *Orthop Clin North Am* **25**: 395, 1994.

119. Whiteside, L. A. and Nagamine, R. Biomechanical aspects of knee replacement design. In W.N. Scott (Ed.), *The Knee*. St. Louis: Mosby, 1994. Pp. 1079-1096.

120. Li, E. and Ritter, M. A. Point - counterpoint of total knee arthroplasty: The case for retention of the posterior cruciate ligament. *J Arthroplasty* **10**:560, 1995.

121. Moilanen, T. and Freeman, M. A. R. Point - counterpoint of total knee arthroplasty: the case for resection of the posterior cruciate ligament. *J Arthroplasty* **10**:564, 1995.

122. Kelly, P. A. and O'Conner, J. J. Why incongruous knee replacements do not fail early. *Trans Europ Orthop Res Soc* **7**:36, 1997.

123. Collier, J. P., Mayor, M. B., McNamara, J. L., Surprenant, V. A., and Jensen, R. E. Analysis of the failure of 122 polyethylene inserts from uncemented tibial knee components. *Clin Orthop* **273**:232, 1991.

124. Feng, E. L., Stulberg, D. S., and Wixson, R. L. Progressive Subluxation and polyethlene wear in total knee replacements with flat articular surfaces. *Clin Orthop* **299**:60, 1994.

125. Blunn, G. W., Joshi, A. B., and Walker, P. S. Performance of ultra-high molecular weight polyethylene in knee replacement. *Trans Orthop Res Soc* **39**:500, 1993.

126. Blunn, G. W., Joshi, A. B., Lilley, A. P., Engelbrecht, E., Ryd, L., Lidgren, L., and Walker, P. S. Polyethylene wear in unicondylar knee prostheses. *Acta Orthop Scand* **63**:247, 1992.

127. Ritter, M. A., Worland, R., Saliski, J., Helphenstine, J. V., Edmondson, K. L., Keating, M. E., Faris, P. M., and Meding, J. B. Flat-on flat, nonconstrained, compression molded polyethylene total knee replacement. *Clin Orthop* **321**:79, 1995.

128. Bartel, D. L., Burstein, A. H., Santavicca, E. A., and Insall, J. N. Performance of the tibial component in total knee replacement. *J Bone Joint Surg* **64-A**:1026, 1982.

129. Wright, T. M. and Bartel, D. L. The Problem of surface damage in polyethylene total knee components. *Clin Orthop* **205**:67, 1986.

130. Engh, G. A., Dwyer, K. A., and Hanes, C. K. Polyethlene wear of metal-backed titial components in total and unicompartmental knee prostheses. *J Bone Joint Surg* **74-B**:9, 1992.

131. Schmalzried, T. P., Scott, D. L., Zahiri, C. A., Dorey, F. J., Sanford, W. M., Kem, L., and Humphrey, W. Variables affecting wear in vivo: analysis of 1,080 hips with computer-assisted technique. *Trans Orthop Res Soc* **23**:275, 1998.

132. Dorr, L. D., Conaty, J. P., Schreiber, R., Mehne, D. K., and Hull, D. Technical factors that influence mechanical loosening of total knee arthroplasty. In L.D. Dorr (Ed.), *The Knee*. Baltimore: University Park Press, 1985. Pp. 121-135.

133. Kilgus, D. J., Moreland, J. R., Finerman, G. A., Funahashi, T. T., and Tipton, J. S. Catastrophic wear of tibial polyethylene inserts. *Clin Orthop* **273**:223, 1991.

134. Wright, T. M., Hood, R. W., and Burstein, A. H. Analysis of material failures. *Orthop Clin North Am* **13**:33, 1982.

135. Eckhoff, D. G., Metzger, R. G., and Vandewalle, M. V. Malrotation associated with implant alignment technique in total knee arthroplasty. *Clin Orthop* **321**:28, 1995.

136. Wasielewski, R. C., Galante, J. O., Leighty, R. M., Natarajan, R. N., and Rosenberg, A. G. Wear patterns on retrieved polyethylene tibia inserts and their relationship to technical considerations during total knee arthroplasty. *Clin Orthop* **299**:31, 1994.

137. Poggie, R., Takeuchi, M., Averill, R., and Nasser, S. Accelerated aging and associated changes in ultra-high molecular weight polyethylene (UHMWPE) microstructure as a function of resin type and consolidation variables. In R.A. Gsell, H.L. Stein, and J.J. Ploskonka (Eds.), *Characterization and Properties of Ultra-High Molecular Weight Polyethylene, ASTM STP 1307*. American Society for Testing and Materials, 1998. Pp. 120-129.

138. Birnkraut, H.W. Synthesis of UHMW-PE. In H.-G. Willert, G.H. Buchhorn and P. Eyerer *Ultra-High Molecular Weight Polyethylene as Biomaterial in Orthopedic Surgery*. Toronto, Hogrefe & Huber Publishers, 1985. Pp.3-5.

139. Schmidt, M.B. and Hamilton, J.V. The effects of calcium stearate on the properties of UHMWPE. *Trans Orthop Res Soc* **21**:224, 1996.

140. Chiba, J., Doyle, J. S., and Noguchi, K. Biochemical and morphological analyses of acticvated human macrophages and fibroblasts by particulate materials. *Trans Orthop Res Soc* **18**:343, 1993.

141. Athanasou, N. A., Heryet, A., Quinn, J., Gatter, K. C., Mason, D. Y., and Mcgee, J. O. Osteoclasts contain macrophage and megakaryocyte antigens. *Journal of Pathology* **150**:239, 1986.

142. Bennet, N. E., Wang, J. T., Manning, C. A., and Goldring, S. A. Activation of human monocyte/macrophages and fibroblasts by metal particles; Release of products with bone resorbing activities. *Trans Orthop Res Soc* **16**:188, 1991.

143. Glant, T. T., Jacobs, J. J., Molnar, G., Shanbhag, A. S., Valyon, M., and Galante, J. O. Bone-resorption activity of particulate-stimulated macrophages. *J Bone Miner Res* **8**:1071, 1993.

144. Gonzales, J. B., Purdon, M. A., and Horowitz, S. M. Response of osteoblasts and macrophages to implant pariculates in a coculture model. *J Bone Miner Res* **10 (Suppl)**:314, 1995.

145. Horowitz, S. M. and Purdon, M. A. Mediator interactions in macrophage particulate bone-resorption. *J Biomed Mater Res* **29**:477, 1995.

146. Maloney, W. J., James, R. E., and Lanesmith, P. Human macrophage response to retrieved titanium alloy particles in vitro. *Clin Orthop* **332**:268, 1996.

147. Maloney, W. J., Smith, R. L., Castro, F., and Schurman, D. J. Fibroblast response to metallic debris in-vitro-enzyme-induction, cell-proliferation, and toxicity. *J Bone Joint Surg* **75-A**:835, 1993.

148. Manlapaz, M., Maloney, W. J., and Smith, R. L. In-vitro activation of human fibroblasts by retieved titanium-alloy wear debris. *J Orthop Res* **14**:465, 1996.

149. Murray, D. W., Rae, T., and Rushton, N. The influence of the surface-energy and roughness of implants on bone-resorption. *J Bone Joint Surg* **71-B**:632, 1989.

150. Pollice, P. F., Silverton, S. F., and Horowitz, S. M. Polymethylmethacrylate-stimulated macrophages increase rat osteoclast precursor recruitment through their effect on osteoblasts in-vitro. *J Orthop Res* **13**:325, 1995.

151. Quinn, J. M., Sabokbar, A., and Athanasou, N. A. Resident and inflammatory tissue macrophages differentiate into osteoclastic bone-resorbing cells. *J Bone Miner Res* **10 (Suppl)**:22, 1995.

Chapter 5

1. Morrison, J B. The mechanics of the knee joint in relation to normal walking. *J Biomech* **3**:51, 1970.
2. Seireg, A. and Arvikar, R. J. The prediction of muscular load sharing and joint forces in the lower extremities during walking. *J Biomech* **8**:89, 1975.
3. Schipplein, O. D. and Andriacchi, T. P. Interaction between active and passive knee stabilizers during level walking. *J Orthop Res* **9**:113, 1991.
4. Andriacchi, T. P., Mikosz, R. P., and Hampton, S. J. Model studies of the stiffness characteristics of the human knee joint. *J Biomech* **16**:23, 1983.
5. Wismans, J., Veldpaus, F., and Janssen, J. A three-dimensional mathematical model of the knee-joint. *J Biomech* **13**:677, 1980.
6. Essinger, J. R., Leyvraz, P. F., and Heegard, J. H. A mathematical model for the evaluation of the behaviour during flexion of condylar type knee prostheses. *J Biomech* **22**:1229, 1989.
7. Blankevoort, L., Huiskes, R., Kuiper, J. H., and Grootenboer, H. J. Articular contact in a three-dimesional model of the knee. *J Biomech* **24**:1019, 1991.
8. Delp, S. L. Surgery Simulation: *A computer graphics system to analyze and design musculoskeletal reconstructions of the lower limb*. Paulo Alto, Ph.D.-Thesis, Stanford University, 1990.
9. Garg, A. and Walker, P. S. Prediction of total knee motion using a three-dimensional computer-graphics model. *J Biomech* **23**:45, 1990.
10. Linn, F. C. and Radin, E. L. Lubrication in animal joints - III. The effect of certain chemical alteration of the cartilage and lubricant. *Arthritis Rheum* **11**:674, 1968.
11. Unsworth, A. Lubrication of human joints. In V. Wright and E.L. Radin (Eds.), *Mechanics of Human Joints: Physiology, Pathophysiology, and Treatment*. New York: Marcel Dekker Inc., 1993. Pp. 137-162.
12. McKellop, H., Clarke, I. C., Markolf, K. L., and Amstutz, H. C. Wear characteristics of UHMW polyethylene: A method for accurately measuring extremely low wear rates. *J Biomed Mater Res* **12**:895, 1978.
13. Davidson, J. A., Mishra, A. K., Poggie, R. A., and Wert, J. J. Sliding friction and UHMWPE wear comparison between cobalt alloy and zirconia surfaces. *Trans Orthop Res Soc* **38**:404, 1992.(Abstract)
14. Soudry, M., Walker, P. S., Reilly, D. T., Kurosawa, H., and Sledge, C. B. Effects of total knee replacement design on femoral-tibial contact conditions. *J Arthroplasty* **1** (1):35, 1986.
15. Fisher, J., Dowson, D., Hamdzah, H., and Lee, H. L. The effect of sliding velocity on the friction and wear of UHMWPE for use in total artificial joints. *Wear* **175**:219, 1994.
16. Dumbleton, J. H. *Tribology of natural and artificial joints - Tribology Series 3*, Amsterdam: Elsevier, 1981.
17. Rullkoetter, P. J. and Hillberry, B. M. Tibio-Femoral contact under dynamic loading. *Trans Orthop Res Soc* **39**:425, 1993.(Abstract)
18. Rabinowicz, E. *Friction and wear of materials*, New York: John Wiley & Sons, 1995.
19. Johnsen, K. L. *Contact Mechanics*, 2nd ed., Cambridge: Cambridge University Press, 1985.
20. Reimpell, J. *Fahrwerktechnik 1*, 2nd ed., Wuerzburg: Vogel-Verlag, 1971.

21. Lloyd, A. I. and Noël, R. E. The effect of counterface surface roughness on the wear of UHMWPE in water and oil-in-water emulsion. *Tribology international* **301**:83, 1988.

22. Andriacchi, T. P., Galante, J. O., and Fermier, R. W. The influence of total knee replacement design on function during walking and stair climbing. *J Bone Joint Surg* **64**:1328, 1982.

23. Draganich, L. F., Andriacchi, T. P., and Andersson, G. B. J. Interaction between intrinsic knee mechanics and the knee extensor mechanism. *J Orthop Res* **5**:539, 1987.

24. Gunsallus, K. L. and Bartel, D. L. Stresses and surface damage in PCA and total condylar polyethylene components. *Trans Orthop Res Soc* **38**:329, 1992.(Abstract)

25. Hood, R. W., Wright, T. M., and Burstein, A. H. Retrieval analysis of total knee prostheses: A method and its application to 48 total condylar prostheses. *J Biomed Mater Res* **17**:829, 1983.

26. Wasielewski, R. C., Galante, J. O., Leighty, R. M., Natarajan, R. N., and Rosenberg, A. G. Wear patterns on retrieved polyethylene tibia inserts and their relationship to technical considerations during total knee arthroplasty. *Clin Orthop* **299**:31, 1994.

27. Blunn, G. W., Joshi, A. B., Lilley, A. P., Engelbrecht, E., Ryd, L., Lidgren, L., and Walker, P. S. Polyethylene wear in unicondylar knee prostheses. *Acta Orthop Scand* **63**:247, 1992.

28. Mahoney, O. M., Noble, P. C., Rhoads, D. D., Alexander, J. W., and Tullos, H. S. Posterior cruciate function following total knee arthroplasty. *J Arthroplasty* **9**:569, 1994.

29. Morrison, J. B. Bioengineering analysis of force actions transmitted by the knee joint. *Biomed Mater Eng* **3**:164, 1968.

30. Draganich, L. F. *The influence of the cruciate ligaments, knee musculature and anatomy on knee joint loading*, Chicago, Ph.D.-Thesis, University of Illinois at Chicago, 1984.

31. Matthews, L. S., Sonstegard, D. A., and Henke, J. A. Load bearing characteristics of the patello-femoral joint. *Acta Orthop Scand* **48**:511, 1977.

32. Van Eijden, T. M., De Boer, W., and Weijs, W. A. The orientation of the distal part of the quadriceps femoris muscle as a function of the knee flexion-extension angle. *J Biomech* **18**:803, 1985.

33. *ADINA Theory and Modeling Guide*, Watertown, Massachusetts: ADINA R&D, Inc., 1987.

34. Beard, B. J. *Origins of wear in the polyethylene component of total knee replacements based on finite element analysis*, Chicago: Thesis (M.S.), University of Illinois at Chicago, 1996.

35. Beard, B. J., Natarajan, R. N., Andriacchi, T. P., and Amirouche, F. M. L. The stress origins of a new striated wear pattern in total knee replacements. *Trans Orthop Res Soc* **21**:465, 1996.(Abstract)

36. DeHeer, D. C. *Stresses in polyethylene components for total knee arthroplasty*, West Lafayette, Indiana Thesis (M.S.), Purdue University, 1992.

37. DeHeer, D. C. and Hillberry, B. M. The effect of thickness and nonlinear material behavior on contact stresses in polyethylene tibial components. *Trans Orthop Res Soc* **38**:327, 1992.(Abstract)

38. Wright, K. W., Dobbs, H. S., and Scales, J. T. Wear studies on prosthetic materials using the pin-on-disk machine. *Biomaterials* **3**:41, 1982.

39. Moilanen, T. and Freeman, M. A. R. Point - counterpoint of total knee arthroplasty: the case for resection of the posterior cruciate ligament. *J Arthroplasty* **10**:564, 1995.

40. Banks, S. A. and Hodge, W. A. Comparison of dynamic TKR kinematics and tracking patterns measured in vivo. *Trans Orthop Res Soc* **40**:665, 1994.

41. Stiehl, J. B., Komistek, R. D., Dennis, D. A., Paxson, R. D., and Hoff, A. A. Fluoroscopic analysis of kinematics after posterior-cruciate-retaining knee arthroplasty. *J Bone Joint Surg* **77-B**:884, 1995.

42. Banks, S. A., Markovich, G. D., and Hodge, W. A. Total knee replacement mechanics during gait. *Trans Orthop Res Soc* **22**:263, 1997.

43. Sathasivam, S. and Walker, P. S. A computer model with surface friction for the prediction of total knee kinematics. *J Biomech* **30**:177, 1997.

44. Walker, P. S., Blunn, G. W., Broome, D. R., Perry, J., Watkins, A., Sathasivam, S., Dewar, M. E., and Paul, J. P. A knee simulating machine for performance evaluation of total knee replacements. *J Biomech* **30**:83, 1997.

45. Hilding, M. B., Ryd, L., Toksvig-Larsen, S., and Stenström, A. Gait affects tibial component fixation. *Trans Europ Orthop Res Soc* **7**:35, 1997.

46. Hilding, M. B., Lanshammer, H., and Ryd, L. Knee joint loading and tibial component loosening. *J Bone Joint Surg* **78-B**:66, 1996.

47. Bartel, D. L., Bicknell, V. L., and Wright, T. M. The effect of conformity, thickness and material on stresses in ultra-high molecular weight components for total joint replacement. *J Bone Joint Surg* **68-A**:1041, 1986.

48. Pruitt, L., Koo, J., Rimnac, C. M., Suresh, S., and Wright, T. M. Cyclic compressive loading results in fatigue cracks in ultra high molecular weight polyethylene. *J Orthop Res* **13**:143, 1995.

49. Schroeder, U., Natarajan, R. N., Andriacchi, T. P., and Wimmer, M. A. The influence of tractive force on the shear stresses in UHMWPE of a TKR component. *Trans Orthop Res Soc* **22**:794, 1997.

50. Little, E. G. Compressive creep behaviour of irradiated ultra high molecular weight polyethylene at 37°C. *Eng Med* **14**:85, 1985.

51. Waldman, S. D. and Bryant, J. T. Compressive stress relaxation behaviour of irradiated ultra high molecular weight polyethylene at 37°C. *J Appl Biomater* **5**:333, 1994.

52. Waldman, S. D. and Bryant, J. T. Dynamic contact stress and rolling resistance model for total knee arthroplasties. *J Biomech Eng* **119**:254, 1997.

53. Blunn, G. W., Walker, P. S., Joshi, A., and Hardinge, K. The dominance of cyclic sliding in producing wear in total knee replacements. *Clin Orthop* **273**:253, 1991.

54. Kelly, P. A. and O'Conner, J. J. Why incongruous knee replacements do not fail early. *Trans Europ Orthop Res Soc* 36, 1997.

55. Li, E. and Ritter, M. A. Point - counterpoint of total knee arthroplasty: The case for retention of the posterior cruciate ligament. *J Arthroplasty* **10**:560, 1995.

56. Sathasivam, S. and Walker, P. S. Optimization of the bearing surface geometry of total knees. *J Biomech* **27**:255, 1994.

57. Szivek, J. A., Cutignola, L., and Volz, R. G. Tibiofemoral contact stress and stress distribution evaluation of total knee arthroplasties. *J Arthroplasty* **10**:480, 1995.

58. Jin, Z. M., Auger, D. D., and Dowson, D. Contact pressure prediction in total knee joint replacements - Part 2: application to the design of total knee joint replacements. *Proc Instn Mech Eng [H]* **209**:9, 1998.

59. Andriacchi, T. P. and Strickland, A. B. Gait analysis as a tool to assess joint kinetics. In N. Berme, A.E. Engin, and K.M. Correia da Silva (Eds.), *Biomechanics of Normal and Pathological Human Articulating Joints*. NATO ASI Series, No. 93, 1985. Pp. 83-102.

60. Perry, J. The mechanics of gait. In V. Wright and E.L. Radin (Eds.), *Mechanics of human joints: physiology, pathophysiology, and treatment.* New York: Marcel Dekker, Inc., 1998. Pp. 83-107.
61. Reeves, E.A., Barton, D.C., Fitzpatrick, D.P., and Fisher, J. A time dependent analysis of cyclic strain accumulation in UHMWPE knee replacements. *Trans Orthop Res Soc* **22**:792, 1997.
62. Estupiñán, J.A., Bartel, D.L., and Wright, T.M. Residual stresses in ultra-high molecular weight polyethylene loaded cyclically by a rigid moving indenter in nonconforming geometries. *J Orthop Res* **16**:80, 1998.

Chapter 6

1. Hood, R. W., Wright, T. M., and Burstein, A. H. Retrieval analysis of total knee prostheses: A method and its application to 48 total condylar prostheses. *J. Biomed. Mater. Res.* **17**:829, 1983.
2. Landy, M. M. and Walker, P. S. Wear of ultra-high-molecular-weight polyethylene components of 90 retrieved knee prostheses. *J Arthroplasty* **Suppl**:73, 1988.
3. Blunn, G. W., Joshi, A. B., and Walker, P. S. Performance of ultra-high molecular weight polyethylene in knee replacement. *Trans Orthop Res Soc* **39**:500, 1993.
4. Blunn, G. W., Joshi, A. B., Hardinge, K., Engelbrecht, E., and Walker, P. S. The effect of bearing conformity on the wear of polyethylene tibial components. *Trans Orthop Res Soc* **38**:357, 1992.
5. Mayor, M. B., Wrona, M., Collier, J. P., and Jensen, R. E. The role of polyethylene quality in the failure of tibial knee components. *Trans Orthop Res Soc* **39**:292, 1993.
6. Collier, J. P., Mayor, M. B., McNamara, J. L., Surprenant, V. A., and Jensen, R. E. Analysis of the failure of 122 polyethylene inserts from uncemented tibial knee components. *Clin Orthop* **273**:232, 1991.
7. Jensen, R. E., Collier, J. P., Mayor, M. B., and Surprenant, V. A. The role of polyethylene uniformity and patient characteristics in the wear of tibial knee components. *Trans Orthop Res Soc* **38**:328, 1992.
8. Wright, T. M., Rimnac, C. M., Faris, P. M., and Bansal, M. Analysis of surface damage in retrieved carbon fiber-reinforced and plain polyethylene tibial components from posterior stabilized total knee replacements. *J Bone Joint Surg* **70-A**:1312, 1988.
9. Wright, T. M., Rimnac, C. M., Stulberg, S. D., Mintz, L., Tsao, A. K., Klein, R. W., and McCrae, C. Wear of polyethylene in total joint replacements: Observations from retrieved PCA knee implants. *Clin Orthop* **276**:126, 1992.
10. Engh, G. A., Dwyer, K. A., and Hanes, C. K. Polyethylene wear of metal-backed tibial components in total and unicompartmental knee prostheses. *J Bone Joint Surg (Br)* **74-B**:9, 1992.
11. Blunn, G. W., Joshi, A. B., Lilley, A. P., Engelbrecht, E., Ryd, L., Lidgren, L., and Walker, P. S. Polyethylene wear in unicondylar knee prostheses. *Acta Orthop Scand* **63**:247, 1992.
12. Wasielewski, R. C., Galante, J. O., Leighty, R. M., Natarajan, R. N., and Rosenberg, A. G. Wear patterns on retrieved polyethylene tibia inserts and their relationship to technical considerations during total knee arthroplasty. *Clin Orthop* **299**:31, 1994.
13. Cornwall, G. B., Bryant, J. T., Hansson, C. M., Rudan, J., Kennedy, L. A., and Cooke, T. D. V. A quantitative technique for reporting surface degradation patterns of UHMWPE components of retrieved total knee replacements. *J Appl Biomater* **6**:9, 1995.
14. Lavernia, C. J., Guzman, J. F., Kabo, M., Krackow, K., and Hungerford, D. Polyethylene wear in autopsy retrieved fully functional totol knee replacement. *Trans Soc Biomat* **20**:82, 1994.

15. Wright, T. M. and Bartel, D. L. The Problem of surface damage in polyethylene total knee components. *Clin Orthop* **205**:67, 1986.

16. Wright, T. M., Hood, R. W., and Burstein, A. H. Analysis of material failures. *Orthop Clin North Am* **13**:33, 1982.

17. McKellop, H. A., Campbell, P., Park, S. -H., Schmalzried, T. P., Grigoris, P., Amstutz, H. C., and Sarmiento, A. The origin of submicron polyethylene wear debris in total hip arthroplasty. *Clin Orthop* **311**:3, 1995.

18. Hutchings, I. M. *Tribology: friction and wear of engineering materials*, London: Edward Arnold, 1992.

19. Rimnac, C. M. and Wright, T. M. Retrieval analysis of knee replacements. In W.N. Scott (Ed.), *The Knee*. St. Louis: Mosby, 1994. Pp. 1251-1260.

20. Andriacchi, T. P., Natarajan, R. N., Wimmer, M. A., Beard, B. J., Karlhuber, M., and Amirouche, F. M. L. Wear patterns in retrieved UHMWPE components of TKR in relationship to loading patterns. In *Characterization and Performance of Articular Surface*. Denver: ASTM Workshop(F04), 1995.

21. Furman, B. D., Ritter, M. A., Perone, J. B., Furman, G. L., and Li, S. Effect of resin type and manufacturing method on UHMWPE oxidation and quality at long aging and implant times. *Trans Orthop Res Soc* **22**:92, 1997.

22. Li, S. and Burstein, A. H. Current concepts review: ultra-high molecular weight polyethylene. *J Bone Joint Surg* **76-A**:1080, 1994.

23. Kanig, G. Ein neues Kontrastierverfahren für die elektronenmikroskopische Untersuchung von Polyäthylen. *Kolloid-Z. u. Z. Polymere* **251**:782, 1973.

24. Smit, T. H., Schneider, E., and Odgaard, A. Star Length Distribution: a volume orientation based concept for the characterisation of structural anisotropy. *J Microscopy* **191**:249, 1998.

25. Pienkowski, D., Jacob, R., Hoglin, D., Saum, K., Kaufer, H., and Nicholls, P. J. Low-voltage scanning electron microscopic imaging of ultrahigh-molecular-weight polyethylene. *J Biomed Mater Res* **29**:1167, 1995.

26. Loos, J. *Kristallation, Morphologie und mechanische Eigenschaften von syndiotaktischem Polypropylen*, Dortmund, Germany: Ph.D. Thesis, Universität Dortmund, 1996.

27. Vezie, D. L., Thomas, E. L., and Adams, W. W. Low-voltage, high-resolution scanning electron microscopy: a new characterization technique for polymer morphology. *Polymer* **36**:1761, 1995.

28. Grood, E. S., Shastri, S., and Hopsen, C. N. Analysis of retrieved implants: Crystallinity changes in ultrahigh molecular weight polyethylene. *J Biomed Mater Res* **16**:399, 1982.

29. Li, S. The identification of defects in ultra high molecular weight polyethylene. *Trans Orthop Res Soc* **19**:587, 1994.

30. Hoechst AG, ®Hostalen GUR. *polymer materials brochure* 1993.

31. Poly Hi Solidur, ®Chirulen. *polymer materials brochure* 1995.

32. Reifer, D. *Untersuchungen an polymeren Oberflächen mit dem Rasterkraftmikroskop*, Münster: Thesis (Diplomarbeit): Physikalisches Institut der Westfälischen Wilhelms-Universität, 1995.

33. McDonald, M. D. and Bloebaum, R. D. Distinguishing wear and creep in clinically retrieved polyethylene inserts. *J Biomed Mater Res* **29**:1, 1995.

34. Cooper, J. R., Dowson, D., Fisher, J., Isaac, G. H., and Wroblewski, B. M. Observations of residual sub-surface shear strain in the ultrahigh molecular weight polyethylene acetabular cups of hip prostheses. *J Materials Science* **5**:52, 1994.

35. Draganich, L. F., Andriacchi, T. P., and Andersson, G. B. J. Interaction between intrinsic knee mechanics and the knee extensor mechanism. *J Orthop Res* **5**:539, 1987.

36. Rose, R. M., Goldfarb, H. V., Ellis, E., and Crugnola, A. M. On the pressure dependence of the wear of ultrahigh molecular weight polyethylene. *Wear* **92**:99, 1983.

37. Blunn, G. W., Walker, P. S., Joshi, A., and Hardinge, K. The dominance of cyclic sliding in producing wear in total knee replacements. *Clin Orthop* **273**:253, 1991.

38. Wright, T. M., Rimnac, C. M., Stulberg, S. D., Mintz, L., Tsao, A. K., Klein, R. W., and McCrae, C. Wear of polyethylene: Observations of retrieved PCA knee implants. *Trans Soc. Biomat.* **17**:247, 1991.

39. Walker, P. S., Blunn, G. W., and Lilley, P. A. Wear testing of materials and surfaces for total knee replacement. *J Biomed Mater Res* **33**:159, 1996.

40. Wang, A., Stark, C., and Dumbleton, J. H. Role of cyclic plastic deformation in the wear of UHMWPE acetabular cups. *J Biomed Mater Res* **29**:619, 1995.

41. Ramamurti, B. S., Bragdon, C. R., O'Conner, D. O., Lowenstein, J. D., Jasty, M., Estock, D. M., and Harris, W. H. Loci of movement of selected points on the femoral head during normal gait. *J Arthroplasty* **11**:845, 1996.

42. Agrawal, C. M., Wirth, M. A., Blanchard, C., and Lankford, J. A study of the wear mechanisms in different wear regimes of UHMWPE. *Trans Orthop Res Soc* **41**:756, 1995.

43. Fischer, A. *Einfluß der Temperatur auf das tribologische Verhalten metallischer Werkstoffe*, Düsseldorf: VDI Verlag, 1994.

44. Davidson, J. A., Poggie, R. A., and Mishra, A. K. Abrasive wear of ceramic, metal, and UHMWPE bearing surfaces from third-body bone, PMMA bone cement, and titanium debris. *Biomed Mater Eng* **4**:213, 1994.

45. Collier, J. P., Mayor, M. B., Suprenant, V. A., Suprenant, H. P., Dauphinais, L. A., and Jensen, R. E. The biomechanical problems of polyethylene as a bearing surface. *Clin Orthop* **261**:107, 1990.

46. Karlhuber, M. *Development of a method for the analysis of the wear of retrieved polyethylene components of total knee arthroplasty*, Chicago/ Hamburg, Thesis (Diplomarbeit), Technical University of Hamburg-Harburg, 1995.

47. Zum Gahr, K. -H. *Microstructure and Wear of Materials*, Amsterdam: Elsevier, Tribology Series Vol.10, 1987.

48. Zhang, B., Zhao, Y., Yang, D., and Wang, E. Scanning tunnelling microscopy on nanofibrils of highly oriented UHMWPE/HDPE films. *J Materials Science* **14**:1275, 1995.

49. Cooper, J. R., Dowson, D., and Fisher, J. Birefringent studies of polyethylene wear specimens and acetabular cups. *Wear* **151**:391, 1991.

50. McNie, C. M., Barton, D. C., Stone, M., and Fisher, J. Stress analysis of microscopic wear of UHMWPE in artificial joints. *Trans Europ Orthop Res Soc* 62, 1997.

51. Fisher, J., Cooper, J. R., Dowson, D., Isaac, G. H., and Wroblewski, B. M. Wear mechanisms and sub-surface failure in UHMWPE acetabular cups. *Trans Orthop Res Soc* **39**:509, 1993.

52. Bristol, R. E., Fitzpatrick, D. C., Brown, T. D., and Callaghan, J. J. Contact stresses on machined vs. molded UHMWPE inserts in total knee arthroplasty. *Proc Am Soc Biomech* **17**:81, 1993.

53. Bristol, R. E., Fitzpatrick, D. C., Brown, T. D., and Callaghan, J. J. Non-uniformity of contact stress on polyethylene insert in total knee arthroplasty. *Clinical Biomechanics* **11**:75, 1996.

54. Wang, A., Sun, D. C., Stark, C., and Dumbleton, J. H. Wear mechanisms of UHMWPE in total joint replacements. *Wear* **181-183**:241, 1995.

55. von Lersner, A. *The comparison of the flow pattern on ultra high molecular weight polyethylene tibial components of total knee replacement in constrained and unconstrained implant designs*, Chicago/ Hamburg, Internship report, Fachhochschule Hamburg, 1996.

56. Seebeck, J. *A method to analyze wear and plastic deformation of retrieved polyethylene components of total knee arthroplasties*, Chicago/ Hamburg: Thesis (Studienarbeit), Technical University of Hamburg-Harburg, 1997.

57. Smit, T.H. *The mechanical significance of the trabecular bone architecture in a human vertebra*, Aachen: Shaker, 1996.

Chapter 7

1. Huang, D. D. and Li, S. Cyclic fatigue behaviors of UHMWPE and enhanced UHMWPE. *Trans Orthop Res Soc* **38**:403, 1992.

2. White, S.E., Whiteside, L.A., McCarthy, D.S., Anthony, M., Poggie, R.A. Simulated knee wear with cobalt chromium and oxidized zirconium knee femoral components. *Clin Orthop* **309**: 176, 1994.

3. Streicher, R.M. and Schön, R. Tribological behaviour of various materials and surfaces against polyethylene. *Tans Soc Biomat* **17**:289, 1991.

4. Treharne, R. W., Young, R. W., and Young, S. R. Wear of artificial joint materials III: Simulation of the knee joint using a computer-controlled system. *Eng Med* **10 (3)**:137, 1981.

5. Blunn, G. W., Walker, P. S., Joshi, A., and Hardinge, K. The dominance of cyclic sliding in producing wear in total knee replacements. *Clin Orthop* **273**:253, 1991.

6. Walker, P. S., Blunn, G. W., Broome, D. R., Perry, J., Watkins, A., Sathasivam, S., Dewar, M. E., and Paul, J. P. A knee simulating machine for performance evaluation of total knee replacements. *J Biomech* **30**:83, 1997.

7. Shaw, J. A. and Murray, D. G. Knee joint simulator. *Clin Orthop* **94**:15, 1973.

8. Mejia, L. C. *Mechanical Testing Systems for Biomaterials/ Biomechanics Research*, Eden Prairie: MTS Systems Corporation, 1994.

9. Davidson, J. A., Mishra, A. K., Poggie, R. A., and Wert, J. J. Sliding friction and UHMWPE wear comparison between cobalt alloy and zirconia surfaces. *Trans Orthop Res Soc* **38**:404, 1992.

10. Stallforth, H. and Ungethüm, M. Die tribologische Testung von Knieendoprothesen. *Biomed Techn* **23**:295, 1978.

11. Dowson, D., Gillis, B. J., and Atkinson, J. R. Penetration of metallic femoral components into polymeric tibial components observed in a knee joint simulator. In Lieng-Huang Lee (Ed.), *Polymer Wear and its Control*. Washington,D.C.: American Chemical Society, 1985. Pp. 215-228.

12. Walker, P. S. and Bullough, P. G. The effect of friction and wear in artificial joints. *Orthop Clin North Am* **4 (2)**:275, 1973.

13. Walker, P. S., Blunn, G. W., and Lilley, P. A. Wear testing of materials and surfaces for total knee replacement. *J Biomed Mater Res* **33**:159, 1996.

14. Blunn, G. W., Lilley, P. A., and Walker, P. S. Variability of the wear of ultra high molecular weight polyethylene in simulated TKR. *Trans Orthop Res Soc* **40**:177, 1994.

15. Walker, P. S., Ben-Dov, M., Askew, M. J., and Pugh, J. The deformation and wear of plastic components in artificial knee joints - an experimental study. *Eng Med* **10**:33, 1981.

16. Engh, G. A., Dwyer, K. A., and Hanes, C. K. Polyethylene wear of metal-backed tibial components in total and unicompartmental knee prostheses. *J Bone Joint Surg (Br)* **74-B**:9, 1992.

17. Rostoker, W., Chao, E. Y. S., and Galante, J. O. The appearances of wear on polyethylene - a comparison of in vivo and in vitro wear surfaces. *J Biomed Mater Res* **12**:317, 1978.

18. Davidson, J. A., Poggie, R. A., and Mishra, A. K. Abrasive wear of ceramic, metal, and UHMWPE bearing surfaces from third-body bone, PMMA bone cement, and titanium debris. *Biomed Mater Eng* **4**:213, 1994.

19. Lampe, F., Grischke, M., Wimmer, M.A., Nassutt, R., Klages, C.P., Hille, E., Schneider, E. Einfluß der Oberflächenhärte der Gelenkkugeln auf abrasive Verschleißmechanismen in der Paarung mit Polyethylen für künstliche Hüftgelenke. *Biomed Techn* **43**: 58, 1998.

20. Nusbaum, H. J. and Rose, R. M. The effects of radiation sterilization on the properties of ultrahigh molecular weight polyethylene. *J Biomed Mater Res* **13**:557, 1979.

21. Bostrom, M. P., Bennett, A. P., Rimnac, C. M., and Wright, T. M. The natural history of ultra high molecular weight polyethylene. *Clin Orthop* **309**:20, 1994.

22. Jahan, M. S., Wang, C., Schwartz, G., and Davidson, J. A. Combined chemical and mechanical effects on free radicals in UHMWPE joints during implantation. *J Biomed Mater Res* **25**:1005, 1991.

23. Waldman, S. D. and Bryant, J. T. Dynamic contact stress and rolling resistance model for total knee arthroplasties. *J Biomech Eng* **119**:254, 1997.

24. Wasielewski, R. C., Galante, J. O., Leighty, R. M., Natarajan, R. N., and Rosenberg, A. G. Wear patterns on retrieved polyethylene tibia inserts and their relationship to technical considerations during total knee arthroplasty. *Clin. Orthop.* **299**:31, 1994.

25. Blunn, G. W., Joshi, A. B., Lilley, A. P., Engelbrecht, E., Ryd, L., Lidgren, L., and Walker, P. S. Polyethylene wear in unicondylar knee prostheses. *Acta Orthop Scand* **63**:247, 1992.

26. Schwenke, T. *Effect of conformity and movement of load on stresses in a total knee component – two-dimensional and three-dimensional finite element study*, Chicago/ Hamburg: Thesis (Diplomarbeit), Technical University of Hamburg-Harburg, 1997.

27. Sellenschloh, K. *Steuerung eines Knieverschleißprüfstandes*, Hamburg: Thesis (Studienarbeit), Fachhochschule Hamburg, 1996.

Chapter 8

1. Stiehl, J. B., Komistek, R. D., Dennis, D. A., Paxson, R. D., and Hoff, A. A. Fluoroscopic analysis of kinematics after posterior-cruciate-retaining knee arthroplasty. *J Bone Joint Surg (Br)* **77-B**:884, 1995.

2. Banks, S. A., Markovich, G. D., and Hodge, W. A. Total knee replacement mechanics during gait. *Trans Orthop Res Soc* **22**:263, 1997.

3. Walker, P. S., Blunn, G. W., Broome, D. R., Perry, J., Watkins, A., Sathasivam, S., Dewar, M. E., and Paul, J. P. A knee simulating machine for performance evaluation of total knee replacements. *J Biomech* **30**:83, 1997.

4. Sathasivam, S. and Walker, P. S. A computer model with surface friction for the prediction of total knee kinematics. *J Biomech* **30**:177, 1997.

5. Davidson, J. A., Mishra, A. K., Poggie, R. A., and Wert, J. J. Sliding friction and UHMWPE wear comparison between cobalt alloy and zirconia surfaces. *Trans Orthop Res Soc* **38**:404, 1992.

6. Soudry, M., Walker, P. S., Reilly, D. T., Kurosawa, H., and Sledge, C. B. Effects of total knee replacement design on femoral-tibial contact conditions. *J Arthroplasty* **1**:35, 1986.

7. Czichos, H. and Habig, K. H. *Tribologie Handbuch - Reibung und Verschleiß*, Braunschweig: Vieweg, 1992.

8. McKellop, H., Clarke, I. C., Markolf, K. L., and Amstutz, H. C. Wear characteristics of UHMW polyethylene: A method for accurately measuring extremely low wear rates. *J Biomed Mater Res* **12**:895, 1978.

9. Wright, K. W., Dobbs, H. S., and Scales, J. T. Wear studies on prosthetic materials using the pin-on-disk machine. *Biomaterials* **3**:41, 1982.

10. Dumbleton, J. H. *Tribology of natural and artificial joints - Tribology Series 3*, Amsterdam: Elsevier, 1981.

11. Fisher, J., Dowson, D., Hamdzah, H., and Lee, H. L. The effect of sliding velocity on the friction and wear of UHMWPE for use in total artificial joints. *Wear* **175**:219, 1994.

12. Moore, D. F. *Principles and Applications of Tribology*, Oxford: Pergamon Press, 1985.

13. Dowson, D. and Jin, Z-M. Micro-elastohydrodynamic lubrication of synovial joints. *Eng Med* **15**:63, 1986.

14. O'Kelly, J., Unsworth, A., Dowson, D., and Wright, V. An experimental study of friction and lubrication in hip prostheses. *Eng Med* **8**:153, 1979.

15. Hall, R. M. and Unsworth, A. Review: friction in hip prostheses. *Biomaterials* **18**:1017, 1997.

16. Knoll, G. Analyse von Druckverteilung und Schmierfilmbildung im künstlichen Hüftgelenk. *Schmiertechnik + Tribologie* **25**:43, 1978.

17. Murakami, T. The lubrication in natural synovial joints and joint prostheses. *JSME International Journal* **33**:465, 1990.

18. Unsworth, A., Pearcy, M. J., White, E. F. T., and White, G. Frictional properties of artificial hip joints. *Eng Med* **17**:101, 1988.

19. Banks, S. A. and Hodge, W. A. Comparison of dynamic TKR kinematics and tracking patterns measured in vivo. *Trans Orthop Res Soc* **40**:665, 1994.

20. Hall, R. M., Unsworth, A., Wroblewski, B. M., and Burgess, I. C. Frictional characterisation of explanted Charnley hip prostheses. *Wear* **175**:159, 1994.

21. Niemann, G. *Maschinenelemente*, Berlin: Springer, 1981.

22. Reinholz, A. *Schlupf-Kraftschlußbeiwert Analyse für Knieendoprothesen*, Hamburg: Thesis (Hausarbeit), Fachhochschule Hamburg, 1997.

Chapter 9

1. Godet, M. The third-body approach: a mechanical view of wear. *Wear* **100**:437, 1984.

2. Bristol, R. E., Fitzpatrick, D. C., Brown, T. D., and Callaghan, J. J. Non-uniformity of contact stress on polyethylene insert in total knee arthroplasty. *Clinical Biomechanics* **11**:75, 1996.

3. MacMillan, N. H. The influence of Poisson's ratio on Hertian contact stresses. *Journal of Materials Science Letters* **8**:340, 1989.

4. Greenwood, J. A. and Williamson, J. B. P. Contact of nominally flat rough surfaces. *Proc Roy Soc Lond* **A295**:300, 1966.

5. Willmann, G. Hüftgelenkersatz – eine tribologische und konstruktive Herausforderung. *Mat Wiss u Werkstofftechnik* **27**: 199, 1996.

6. Way, S. Pitting due to rolling contact. *J Appl Mech* **2**:A49, 1935.

7. Wang, A., Stark, C., and Dumbleton, J. H. Role of cyclic plastic deformation in the wear of UHMWPE acetabular cups. *J Biomed Mater Res* **29**:619, 1995.

8. McNie, C. M., Barton, D. C., Stone, M. H., and Fisher, J. Prediction of plastic strains in ultra-high molecular weight polyethylene due to microscopic asperity interactions during sliding wear. *Proc Inst Mech Eng [H]* **212**:49, 1998.

9. Boresi, A. P. and Sidebottom, O. M. *Advanced Mechanics of Materials*, 4th ed., New York: John Wiley & Sons, 1985.

10. Premnath, V., Harris, W. H., Jasty, M., and Merrill, E. W. Gamma sterilization of UHMWPE articular implants: an analysis of the oxidatin problem. *Biomaterials* **17**:1741, 1996.

11. Muratoglu, O. K., Bragdon, C. R., Jasty, M., and Harris, W. H. An oxidation mechanism in gamma sterilized ultra-high-molecular-weight-polyethylene components. *Trans Europ Orthop Res Soc* **10**:184, 1996.

Medical Terminology

abduction	movement of a limb away from the median line (compare to adduction)
adduction	movement of a limb toward the median line (compare to abduction)
antagonistic	muscle activity characterized by mutual opposition in action
anterior	denoting the front side of the body (compare to posterior)
arthroplasty	surgical procedure to restore the function of a diseased joint (*jargon*: artificial joint)
aseptic	free of living pathogenic organisms
bone cement	polymethylmethacrylat (PMMA) used to fix implants to the bone
capsulectomy	removal of the anatomic joint capsule
cartilage	an elastic tissue occurring at the articulating ends of bone
condyle	a rounded articular surface at the extremity of a bone
cortical bone	the outer layer of bone characterized by a higher density and rigidity relative to cancelleous bone
cystic lesion	wound relating to a cyst
cytokine	any of various proteins, secreted by cells, that carry signals to neighbouring cells
debris	fragments or remnants of something destroyed or broken, wear particles
distal	situated away from the center of the body (compare to proximal)
extensor apparatus	the tissue structures responsible for leg extension;
femoral	relating to the femur or thigh
femur	the bone of the thigh
fibula	the thinner of the two bones of the shank (splint bone)
frontal plane	plane in the medio-lateral direction
granuloma	a tumor composed of granulation tissue
hemi-	*prefix* signifying one half
immunological	characterized by the ability of an organism to resist disease
implant	a metallic or plastic device employed in joint reconstruction
in situ	in the body
in vivo	in living cell, organs or animals
in vitro	in laboratory devices

lateral	on the side; farther from the median (compare to medial)
ligament	band of tough fibrous tissue connecting bony structures
macrophage	any large phagocytic cell occurring in the blood, lymph, and connective tissue
malalignment	displacement
medial	nearer to the median (compare to lateral)
meniscus	fibrous cartilage between the bones at the knee joint
monocyte	specific type of macrophage
myoelectric	relating to the electrical properties of muscle
osteoblast	cell associated with bone formation
osteoclast	cell associated with bone absorption
osteolysis	absorption and destruction of bony tissue
patella	kneecap
peri-	*prefix* denoting around
phagocytosis	process by which a cell, such as macrophage, ingests microorganisms, other cells, and foreign particles
posterior	denoting the back side of the body (compare to anterior)
proximal	situated close to the center of the body (compare to distal)
resection	excision of a part of a bone
revision	surgical procedure by which an implant is substituted
sagittal plane	plane in the antero-posterior direction
synergistic	coordinated or correlated action
synovial fluid	lubricating fluid, secreted by the joints
tendon	fibrous cord or band that connects a muscle to a bone or other structure
tibia	the medial and bigger of the two bones in the shank (shinbone)
tibial plateau	proximal extremity of the tibia
transverse plane	plane lying across the long axis of the body
varus	joint twisted outward (compare to varus)
valgus	joint twisted inward (compare to valgus)

Curriculum Vitae

Markus Wimmer was born on December 17, 1965 in Munich, Germany. After graduating at the Christoph Scheiner Gymnasium, Ingolstadt (diploma: Abitur), he attended the Technical University of Munich in 1986 to study mechanical engineering. Specializing in construction and development he developed a special interest in mechanics and, in particular, sports related biomechanics. This resulted in part time positions as student tutor at the Institute of Mechanics (Prof. Lippmann) and sports instructor at the Institute of Sports. As a collaborator within the Scientific Service Team of the University of Cologne and Audi (program sponsor), he was involved in skiing mechanics and materials testing for the German skiing national team. The development of a ski binding for the measurement of ground reaction forces during downhill racing led to his graduation (Diplom-Ingenieur, univ.) in 1992.

During a short period as research assistant at the Technical University of Hamburg-Harburg (1992-1993), Markus Wimmer moved his interest to the field of clinical biomechanics and started working on wear phenomena under the supervision of Prof. Erich Schneider. In September 1993 he was awarded with a scholarship from the German Academic Exchange Office to spent one year of research at the Department of Orthopedics at the Rush Presbyterian St. Luke's Medical Center, Chicago in USA. There, Prof. Thomas Andriacchi taught him the principles of knee biomechanics and gait analysis. In September 1994, he followed the invitation of Prof. Schneider to support him in establishing implant tribology as one of the main topics of research at the Biomechanics Section in Hamburg. Several projects were started, including the evaluation of polyethylene wear at the tibial component of artificial knee joints. To account for the interdisciplinary character of this project, intensive collaborations to Prof. Andriacchi (knee biomechanics, Stanford University), Prof. Petermann (polymers, University of Dortmund), and Prof. Fischer (tribology, University of Essen) were established and resulted in this dissertation. Since October 1997, Markus Wimmer is affiliated with the AO Research Institute at Davos, Switzerland, and is heading the Joint Replacement Group.